TEN ENGINEERS
WHO MADE
BRITAIN
GREAT

THE MEN BEHIND THE
INDUSTRIAL REVOLUTION

ANTHONY BURTON

The
History
Press

Cover illustrations: Front, clockwise from top: Galton Bridge, built by Thomas Telford in Smethwick (British Waterways); British flag (Shutterstock); Isambard Kingdom Brunel's *Great Eastern* (George Stacey); Richard Trevithick's portable steam engine (Francis Trevithick, *A Life of Richard Trevithick*, 1872). *Back, clockwise from top left:* John Metcalf; Richard Trevithick; Thomas Telford; Robert Stephenson; John Smeaton; George Stephenson.

First published 2023

The History Press
97 St George's Place, Cheltenham,
Gloucestershire, GL50 3QB
www.thehistorypress.co.uk

© Anthony Burton, 2023

British Library Cataloguing in Publication Data.
A catalogue record for this book is available from the British Library.

ISBN 978 1 80399 112 2

Typesetting and origination by The History Press
Printed and bound in Great Britain by TJ Books Limited, Padstow, Cornwall.

MIX
Paper from
responsible sources
FSC® C013056

Trees for Life

CONTENTS

PREFACE

The idea for this book came from reading Samuel Smiles's famous work of 1862, *Lives of the Engineers*, in which he portrayed them as heroic figures. It has always been valued for its historic role in treating the subject in a way that made the work both popular and accessible to non-specialists. But there are problems. Smiles likes to have his heroes face and overcome obstacles, some of which research suggests are inventions or very twisted versions of actual events, and he also includes dialogue that no one could ever have recorded and written down at the time. What follows is perhaps a little more sober, but it is hoped that in writing about these ten great men, the readers will appreciate just what they achieved; achievements that need no embellishments.

I

JOHN METCALF

The story of John Metcalf, popularly known as Blind Jack of Knaresborough, is a remarkable one. His exploits can seem almost too extraordinary to be true, but almost everything we know about him comes from the autobiography that he dictated to a publisher in York. He can certainly never be accused of modesty and there are times when readers may feel twinges of doubt creeping in. What is not in question is that he was a remarkable man and that would still be the case if only half the stories were true.

Knaresborough is a Yorkshire market town that stands on a hill, rising steeply above the River Nidd, and dominated by the ruins of a Norman castle. Metcalf was born here near to the castle to a working family on 15 August 1717. He was sent off to school at the age of 4, but two years later he contracted smallpox, which left him completely blind. However devastating this must have been for a young boy, he proved a remarkably resilient character, for within just six months he was setting out to go to the end of the street and return home without a guide. Soon he was able

Image: John Metcalf. (*The Life of John Metcalf*, 1795)

to wander all over the town and began mixing as an equal with boys of his own age. He did what the others did – climbed trees and went for walks in the surrounding countryside. Although his parents were simply described as 'working people' they were obviously not poor, as the father kept horses and encouraged his son to take up riding. The boy revelled in this new achievement and was encouraged by a local gentleman called Woodburn, who was master of a pack of hounds and took the lad, now generally known as Jack, with him on the hunt. From the start he was not only able to cope remarkably well with his blindness, but showed a certain reckless bravado that was to be the hallmark for much of his life.

Metcalf's boyhood was marked by a series of episodes that showed him to be up for any challenge – not all of which ended well. He was known to have 'borrowed' one or two of Woodburn's hounds without permission and gone off on expeditions on his own – and those lasted until the hounds were found to have attacked a farmer's lambs. He and his friends were enthusiastic visitors to a plum tree and Jack was the one who was always given the job of climbing and chucking the fruit down to the others. All went well until the boys on the ground saw the owner approaching and promptly ran off, leaving Jack stranded. He clambered down, but while following the others he tumbled into a gravel pit and gashed his face: plum stealing joined illegal hunting in the list of abandoned activities. Another of his passions was swimming in the river, where he was the most boisterous of the group – often pulling the others down and swimming over their heads. The picture is of a strong, fun-loving boy who would have been quite unusual even without his blindness, but who was determined not to let the disability spoil his enjoyment of life.

Metcalf was growing up to be a strong and powerful young man – he would eventually grow to be over 6ft tall. But the mischievous spirit did not seem to leave him in his later teens. He was visiting Scriven, a village a mile or so from Knaresborough, where there was a dispute going on about some sheep penned up outside the inn. Metcalf left the two men arguing, and jumped over the wall into the pen, where he started catching sheep and depositing them over the wall. It was easy at first as the sheep were huddled together, but as more went over the wall, the rest became more and more difficult to catch. Anyone who had happened to pass by at the time would have been met with the amazing vision of a young blind man chasing a bleating sheep around an enclosure. But he got them all over the wall and went home. When the two men left the inn, they faced an even more astounding and puzzling sight: a flock of sheep grazing outside the pen that had been, and still was, locked.

The time came when decisions had to be made about Metcalf's future. There is a long tradition behind one occupation for the blind – playing the fiddle. The boy set about learning the violin, playing by ear as there was no possibility of reading music. He soon became sufficiently competent to perform at country dances at Knaresborough and nearby towns. He continued to enjoy hunting and also took an interest in cock fighting – and he developed a growing reputation for devilry and pranks. As this is an account that should be centred on his future role as an engineer, it is not necessary to repeat every story that has been told about him, but one at least shows the resourcefulness and initiative that were to be a feature of everything he tackled.

In 1732 he finally got regular employment at the fashionable Queen's Head hotel in Harrogate, which had a long room for dancing. The previous fiddler had finally decided to retire, not unreasonably, having reached his 100th birthday. Metcalf proved very successful, and he received another invitation, this time to the Green Dragon also in Harrogate. He seems to have made himself extremely popular with the gentry, for the Squire of Middlethorpe near York invited him to stay with him for the winter, where he could go out with him and his hounds and practise music. He also arranged for a more accomplished musician to give him lessons to improve his skills. He was returning on horseback through York when the landlord of the George Inn, who knew him, asked if he was going back to Knaresborough as he had a guest who needed a guide to Harrogate. Metcalf agreed to take the stranger along, provided he didn't mention his blindness. It was only when they had safely reached their destination that the stranger discovered, to his horror, that he had been led all the way by a blind man.

Metcalf returned to his regular playing in Harrogate, and one of the inns he visited was the Royal Oak, where the landlord's daughter, Dorothy Benson, took a liking to the young man, and they began a secret liaison. During this time, there is a whole range of stories about him and his exploits, in particular a readiness to bet on anything, from his prowess on a horse to his skill at cards. At this point, one does begin to stop and wonder how accurate all these tales might be. It is just about acceptable to find a blind man galloping across the countryside, but how does he possibly play card games? Did he have a friend whispering in his ear to tell him what he held and what was going on? Or did he perhaps retain a slight amount of vision, enough to make out rough shapes but no details? Total blindness seems to have been accepted by all who knew him, so it is a puzzle that may never be solved.

By the time he reached the age of 21, Metcalf was thoroughly enjoying life, and quite happy to take on any new pleasure that came his way. Two of his friends had a sister who shared his attitude, and she suggested that if he put a candle in his window, she'd come that night for a visit. The candle was lit, the visit was made and a few months later the consequences appeared. The young woman insisted he married her, but he seems to have thought that, as she had instigated the whole affair, he had no moral obligation – and he was still in love with Dorothy Benson. Benson, it seems, was remarkably understanding, and simply told him that he definitely should not marry the young woman. A law officer soon appeared on the scene demanding that he do his duty, but Metcalf offered to pay the girl off – and it was agreed that a sum of £30 to £40 would seal the matter. He told the officer to wait while he went to Harrogate to get his money: the officer waited, but Metcalf never appeared. He was on his way to Scarborough, where he stayed for several months, before travelling to Whitby and taking a boat to London.

In the capital, he met a fellow northerner, who played the pipes and who introduced him to Colonel Liddell, the Member of Parliament for Berwick-upon-Tweed, who lived near Newcastle, but spent three weeks of the year at Harrogate. He was about to make the trip north, on horseback with his servants, and invited Metcalf to join him. But Metcalf declined, saying he would walk as it would be quicker. They agreed to meet up each evening along the way. Metcalf being quicker probably seemed unlikely to the colonel, but it says a great deal about the state of the roads of Britain at that time that he was usually the first to arrive at each stage of the journey – and also goes some way to explain why Metcalf appears in this book at all.

It has often been noted, that at the start of the eighteenth century, the roads of Britain were in a far worse condition than they had been under the Romans. In medieval times, the King's Highway was often little more than a well-established track. This was not a problem when the roads were mostly used by the poor on foot or the rich, either on horseback or being carried in litters. Transport of goods was mainly in broad-wheeled wagons. If the middle of the road got boggy after rain, travellers simply moved out to the drier edges, spreading the highway, which was not a problem until the nature of wheeled traffic changed. But in 1555, the Duke of Rutland was reported to have been seen riding in a carriage suspended in a narrow-wheeled frame. When Queen Elizabeth I acquired one of the new-fangled devices, the trend was set – anyone who could afford a carriage wanted one. They also wanted a smooth ride on a decent surface. The big question was – who would pay for better roads?

The government attempted to legislate better roads into existence by passing a series of laws covering such matters as wheel widths, which were almost entirely ineffective. The problem lay with the old system that had existed since the Middle Ages that made local authorities responsible for the highways within their parishes. A surveyor of roads was appointed by the parish to supervise road maintenance, and he was empowered to call on the local men to carry out what was known as their Statute Duty to spend six days a year on the job. Not surprisingly, the idea was seldom greeted with any enthusiasm. John Hawkins described the situation in his *Observations on the State of the Highways* published in 1763:

> Let us now see in what Manner the Law at present under Consideration is observed in those few Parishes, where the Inhabitants are disposed to yield obedience to the Letter of it: the Days for performing the Statute Duty are so far from being considered as Days of Labour, that as well the Farmers as the common Day-Labourers, have long been used to look on them as Holidays, as a kind of Recess from their accustomed Labour, and devoted to Idleness and in concomitant Indulgences of Riot and Drunkenness.

The surveyor did not have to use the parishioners, but could, if the parish provided the money, pay day labourers to do the work. But the money was seldom forthcoming, and if it did arrive it was scarcely adequate for the task. The problem was compounded by the fact that the surveyor usually had no training of any sort and no experience of road construction. All too often he bumbled along with an unwilling workforce, or as Hawkins aptly described the situation: 'A Contest between Ignorance armed with Authority on the one Side, and invincible Obstinacy on the other.' In the circumstances, it is scarcely surprising that the roads were often in a terrible condition. Daniel Defoe, travelling through Britain at the start of the eighteenth century, noted that the road between Dunstable and Nottingham was so bad, with a surface that was little more than soft clay, that horses were known to drop dead with the strain of trying to pull carts along it. While in the Stoke-on-Trent area, potholes had a literal meaning, as local workers were quite likely to take clay from the highway for their works. This was not a matter of scooping up a little to make a teapot, but removing such serious quantities that on one occasion a traveller at night fell into one of these holes and drowned. Small wonder then that when Metcalf and the colonel set out for Harrogate, it was the walker who arrived each day before the riders.

The situation was, however, beginning to change. It was clear that neither the government not the local authorities were going to make matters better. The answer was to turn to private contractors. They would build the roads and recoup the expense by charging those who used them: these new toll roads became known as turnpikes, from the gates that crossed them where the money was collected. It was on these new turnpikes that Metcalf was to make his mark, but there was a great deal to happen to him before then.

Back in Harrogate, he once again made his way to the Royal Oak and Dorothy. He returned to his old life and was soon a regular performer at Ripon. He began to think seriously about marriage and started saving, leaving his money with Dorothy, who was more likely to look after it well than he was. But old habits soon took over. He had saved £15, but took it from Dorothy as an 'investment', which turned out to be a cock fight – and lost the lot. The remainder went on a horse, which won, but Dorothy must by now have been wondering whether her intended would ever settle down. In the meantime, she had another suitor and to Metcalf's chagrin he heard the banns being read in the parish church.

Metcalf was forced to take decisive action and propose to Dorothy – on the eve of her wedding. It seems that little in his life followed a conventional path. That night he 'borrowed' a horse. Dorothy hastily packed a few clothes and off they went, where she was left with a friend some miles from Harrogate. Metcalf then galloped back to Harrogate to be there when it was discovered that the bridegroom no longer had his bride. He pretended to know nothing of it, but the story soon came out: the couple were married and set up house in Knaresborough. He continued to make money by playing his violin at Harrogate, but he also bought four-horse and one-horse chaises that he rented out to the public. The business was successful for a while, but as competition grew with Harrogate's increasing popularity as a spa he found it difficult to make a good profit, so he began a brand new business, travelling to the coast for fish, which he brought back on four pack horses to sell in Leeds and Manchester. He had become an experienced traveller and knew the roads of the north – and their shortcomings – well. He often found himself crossing streams and rivers where there were no bridges, and one winter on his way to Leeds the ice gave way and one of his horses fell in. He unloaded the animal to free it, but as soon as it was out the beast decided it had had enough and bolted back to Knaresborough. He managed to distribute his load over the remaining animals, but it was a miserable experience. He eventually found that being a fishmonger entailed a great deal of unpleasantness that was, unfortunately, not accompanied by a suitable profit, so he returned to earning a living at Harrogate with his fiddle.

His life changed once again, very dramatically in 1745, when Bonnie Prince Charlie raised a Highland regiment to march on England to reclaim the throne for the House of Stuart. William Thornton was a wealthy man and an enthusiastic patriot. He proposed at a meeting at York that money raised to support the regular army should be used to fund a regiment of 4,000 volunteers to join the cause. The idea was not well received, so he decided to raise a company of his own and at his own expense. Thornton visited Knaresborough, where he gave a blood-curdling speech, describing what would happen if the Scots and their French allies invaded England: 'if not vigorously opposed they would violate all our wives, daughters and sisters'. Metcalf, never one to refuse a challenge, promptly volunteered to join Thornton, who had appointed himself to the rank of colonel, and was invited to join the recruiting sergeant in encouraging others to sign up. Fearing the threat of seeing the females in their families being raped might not be enough, Thornton promised them promotion in the army and lucrative contracts from the government once the battle was won. With such impossible bribes, he persuaded 140 men to volunteer, of whom sixty-four were selected as privates in the company. He got his men blue tunics from tailors in Leeds and with arms provided from London began drilling his recruits. They then went to the colonel's home at Thornbury for more drill, and it seems a rather enjoyable time, feasting on roast oxen and being served 'seven years' old beer' – this was considered a great luxury, although it is hard to imagine what it would have been like.

The day came when they were ready for action, and the company marched off to Newcastle, encouraged by Metcalf playing patriotic airs and marches. Once they arrived, they made camp on Newcastle Moor, where they joined General Wade's army. Thornton bought tents for the men, each being supplied with a blanket, and a marquee for himself. After staying there a week, they were ordered to Hexham and began a long march in atrocious conditions, with hail and snow. They were severely hampered by the wretched state of the roads, and the engineers had to carve out a way for the baggage train and artillery, so they took fifteen hours to cover just 7 miles. Straw had been laid out for them at the Hexham site, but the ground was so hard the men could not drive in the tent pegs. Thornton had gone into Hexham, so Metcalf took it upon himself to take control, and told the men to burn the straw to get warm. He played his fiddle while they danced around the fire.

The whole character of the war was changing: the Jacobites had advanced as far as Derby, but there they were turned back by forces under the command of the Duke of Cumberland. The Scots began a retreat back to Scotland, pursued by the English, who were once again hampered by the poor state of the roads. Eventually the Highlanders finished up at the hamlet of Torwood between Falkirk and Stirling, where they camped. The English camp, which included Thornton's company, was just 3 miles away. An attack was expected, but the Scots were apparently moving away towards Stirling. It is a curiosity of this campaign that the English were so often led by aristocracy, who seemed unwilling to put aside social engagements. General Hawley, in charge of the English, happy that things were going well, went off to Callander House for breakfast with Lady Kilmarnock, whose husband was fighting with the Scots.

While the general was supping his morning tea, the English found to their horror that, far from retreating, the Scots had turned and were about to attack. Thornton's men were ordered to join the matrosses, the soldiers who assisted the gunners. When the cannon sank in the boggy ground, Thornton rode off to join the cavalry attacking the Scots line. The attack was repulsed, and the English were ordered to make a stand at Falkirk. It was then discovered that the rain had soaked the powder, and there was little chance of holding the position. A retreat was ordered to Linlithgow, which would be easier to defend. No one, however, had realised, just how close the Scots were. Twenty of Thornton's men were captured, together with the lieutenant. Thornton himself had gone to a nearby house to change out of his wet clothes and was still there when the rebels came to the house. The lady of the house hid him in a closet, and put a dresser in front of the door. The house was full of Highlanders, but the woman managed to sneak food to the unfortunate Thornton. It was a desperately uncomfortable position, but the woman arranged to get help from an old loyalist carpenter, and the following night they dressed Thornton up in plaid and brogues, put a black wig on his head and gave him a bag of carpenters' tools. Before they left, Thornton gave the kind woman 8 guineas from the 10 guineas he still had on him. In disguise they made their way through the crowds of Scots and reached Edinburgh and the English army in safety.

Metcalf knew none of this adventure, and if that story reads like something out of a book by Robert Louis Stevenson, then the next narrative is, if anything, even more dramatic. He remembered that Thornton had two horses stabled nearby and went to fetch them. They were both saddled and as he led one out was stopped by a band of Scots, who demanded the horse. When he refused, they drew pistols and said it was needed by Prince Charles, and Metcalf handed it over, pretending to be a loyal supporter of the prince. He eventually made his way to Edinburgh, still not knowing what had happened to Thornton. He was determined to locate him, but he found a man he knew from Knaresborough, who supported the Jacobites, and told him he was now wanted to play fiddle for Prince Charles, which was convincing as the general view in Scotland was that the English were losing the war. Metcalf was offered the help of an Irish spy, who would help him cross the English lines and make his way to Falkirk. They were, in fact, stopped by the English, but Metcalf insisted on going ahead with his plan and eventually reached Falkirk. He was well received there and, when questioned by the Scots who had taken the town, told them he was a fiddler employed by the English officers, but his ambition was to play for Prince Charles. However, someone recognised him as the fiddler from Knaresborough and he was arrested and imprisoned for three days. He was court-martialled, but stuck to his story and was released.

He eventually made his way to Edinburgh, where he and Thornton returned to the army. They continued with it as the campaign moved north, and were present at Culloden, where the rebel army suffered its final bloody defeat. Metcalf was now free to return home.

He returned to his old habits, playing for money at Harrogate, but his travels had convinced him that there was money to be made in trading goods from Scotland. He bought worsted and Aberdeen stockings, which he sold in York at a profit. He bought horses that he took north of the border and returned with Galloway cattle. He also discovered that an even more lucrative trade was possible: he became a smuggler. He was nearly caught out by exciseman at Chester-le-Street, where he had 100cwt of tea hidden under a load on pack animals, but talked his way out of it without being discovered. His biggest scheme involved selling a string of horses in Scotland and using the money to buy £200 worth of tea, rum and brandy. He put his goods on a vessel bound for Leith, and walked there himself, expecting it to be there more or less the same time as the ship. He had an anxious time as the vessel failed to appear – six weeks went by before it finally berthed and Metcalf boarded it for Newcastle. His relief was short lived. A storm blew up, the mate was washed overboard and the mainsail blew away. The ship seemed doomed to founder off the Norwegian coast, but the captain managed to turn the vessel and head back to Scotland hoping to reach Arbroath. No pilot was willing to help the vessel into harbour, but somehow it made it, but not without hitting the harbour wall and taking in water. When she was finally made safe, there was found to be 5ft of water in the hold. Metcalf set off again in another vessel with his illicit cargo – he had 400 gallons of run and brandy for which he had permits and another 40 he was hoping to smuggle in. The ruse was discovered, but once again he managed to wriggle out of the difficulty. He explained that the 400 gallons were for sale and quite legitimate, while he had intended the remainder for treating the sailors and to take home for his personal use. The story was believed, otherwise he would have forfeited the whole lot.

In 1751 Metcalf began yet another new venture, running a stage wagon between Knaresborough and York, a journey of roughly 21 miles each way. This would have been an ordinary wagon with a canvas cover pulled over metal hoops and would have been used to carry both goods and passengers. He made two return trips a week in summer and one a week in winter. That same year, an Act was passed for a turnpike road to be constructed from Harrogate to Boroughbridge. Metcalf offered to take on the construction of 3 miles of the new road between Minskip and Frensby (now Ferrensby). There is little indication of what was involved in the way of laying a foundation, but the top surface was to be gravel. All the material was to be available from one gravel pit, and Metcalf recruited a team of men, who were housed locally, and built temporary stabling for a dozen horses. This was his first venture into road construction, but it marked the start of the period of his life that earned him his place in history – and, indeed, in this book.

Being blind, Metcalf had developed his own system of mental arithmetic and used it not merely for everyday business, but for practical measurements as well. According to the biography, plans were being put forward for a new bridge at Boroughbridge and a general meeting was called to receive estimates. Various masons put in bids, at a great variety of prices, and with various estimates of the amount of stone needed for the job. At this point Metcalf offered to take on the work, though he had no experience of bridge building whatsoever. When asked to explain his ideas, he had his reply ready:

> The span of the arch, 18 feet, being a semi-circle, makes 27; the archstone must be a foot deep, which if multiplied by 27, will be 486, and the base will be 72 foot more – This for the arch: it will require good backing, for which purpose there are proper stones in the old Roman wall at Aldborough, which may be brought, if you please to give directions to that effect.

There is a puzzle here, because no new bridge was built here at that time. However, the old bridge dated back to 1562 and could well have been in such a dilapidated condition that it required extensive rebuilding. He also refers to the span of the arch, suggesting there was just one, but as one can see from the photograph (p. 20), there are three arches. This tends to reinforce the idea that he was actually repairing a part of the bridge, not constructing a brand new one.

He was soon involved in a number of road constructions, starting with the route from Knaresborough bridge to Harrogate. Along the way, when investigating the route, he told the men with him that the grass he was walking on seemed different and he thought there was stone

1.1 The bridge at Boroughbridge, the construction of which Metcalf was partly responsible. (John D. Hall)

enough. The turfs were lifted and revealed an older stone causeway, which Metcalf decided was probably Roman. Today, this would be considered an important archaeological find: then, it was simply a useful source of stone for the new road. Faced by a patch of boggy ground, the surveyor in charge of operations thought it would be necessary to go around it, but Metcalf thought otherwise. The surveyor agreed that if he could make a sound road over the bog, he would pay the same amount in terms of distance as though he had gone around. Metcalf used bundles of gorse and heather to soak up the water and to form a firm base for his road – a technique he would use again in the Pennines.

He would go on to build many more roads in the north. In surveying the route, he used a viameter, a large diameter wheel with a dial on the handle. As he walked along, the wheel turned at his side and at each full revolution the hand on the dial moved on. The dial was calibrated so that he could read the distances by touch. In his travels around Britain, Metcalf had all too often been held up by roads made impassable by rain, churning the surface into a cloying mass. To remedy this, he ensured his roads had sound foundations and the top surface was curved so that water ran off down to the sides. This was not an original idea: as early as the sixteenth century the famous Italian architect Andrea Palladio had designed similar roads, with good bases and pronounced cambers. But, until the turnpike age made it profitable, road builders in Britain had not felt the need to go to that much trouble and expense. There is, therefore, a very good case for describing Blind Jack as Britain's first modern professional road builder. He was not quite alone in constructing sound roads. Following the end of the Jacobite rebellion, where the English army had been frequently held up by poor communications, a series of military roads were built in Scotland. One extensive section of a road built under the direction of General Wade can be seen crossing Rannoch Moor. The surface is too worn to be able to see clearly whether there was an extensive camber, but the drainage ditches to either side suggest that at least the engineers responsible were aware of the need to remove surface water. That others had the same idea does not distract from Metcalf's achievements.

1.2 A statue of Metcalf with his viameter in the Market Square at Knaresborough. (M. Taylor)

1.3 An illustration by the eighteenth-century Artist W.H. Pyne showing typical road construction of the time. (From W.H. Pyne, *Microcosm*, 1807)

Among the most demanding routes he undertook was a road down the Colne Valley from Huddersfield via Marsden to a point just outside Oldham. The initial survey had included a section over Pole Common that included an extensive area of marsh. Metcalf was told he should excavate the area until he reached a solid bottom. Not only would that have involved vast excavation work, as soundings suggested that they would have to dig down to a depth of 9ft, but the road would have to dip down into the hollow, which was likely to fill up with snow every winter. He returned to the idea first tried on the Knaresborough–Harrogate road. He had the men make up bundles of heather in squares, adding another layer of bundles on top and pressing them well down and afterwards laying on the stones and the top layer of gravel. The system worked and it was said that the road only needed to be repaired for the first time after twelve years' use. The next obstacle was to take the road across the wasteland of Standedge Moor and down the far side into Lancashire. The line of the road is more or less that of the modern A62.

The narrative is interrupted in the biography with the first mention of his wife, Dorothy, of whom almost nothing has been written since their elopement. In fact, they had had four daughters. While Metcalf was working on a new road in Cheshire, Dorothy became ill. He brought her to Stockport where one of his married daughters lived and where

1.4 Paying tolls on a turnpike road, also by Pyne. (From W.H. Pyne, *Microcosm*, 1807)

there was a doctor, who was said to be an expert in curing rheumatic complaints. He could do nothing and Dorothy died at Stockport in the summer of 1778 at the age of 60. She seems to have been a lady of modest needs. Metcalf did his best to provide her with gifts and comforts, but she wanted and needed little.

As always with Metcalf, when something new turned up that looked as if it might show a profit, he was keen to get his share. His son-in-law manufactured stockings in Stockport, and Metcalf decided there was money to be made in the textile industry. The cotton industry was booming, and the spinning process had been first mechanised by James Hargreaves, who patented his jenny in 1770. It was manually operated, but one worker could drive a whole series of spindles to produce yarn. Before spinning, the cotton fibres had to be aligned, a process known as carding, because originally it involved pulling the threads through a card studded with metal spikes. That process too had been mechanised. Metcalf bought a carding engine and six jennies and began manufacturing.

Unfortunately, he found the price he got for his yarn meant he was working at a loss – the first water-powered cotton mills were already coming into use. He was not a man to be easily discouraged, so he bought looms and began making cloth. He produced calico, which he had bleached and printed, and velveret, an inferior form of velvet, with a cotton base and a silk pile. He sent off 800 yards of material for sale in Yorkshire and, having got the business up and running, left his son-in-law in charge and began to make his way back to Yorkshire.

On his way home, he stopped at Marsden, where he heard of a road and bridge being needed and promptly bid for both. He also began work on a major project in Lancashire, linking Bury and Accrington, with a branch to Blackburn. The terrain was far from ideal for road construction, a succession of hills and hollows, so that in some places the road had to be carried on embankments as much as 30ft high, while in others it had to lie in deep cuttings. He was paid £3,500 for the work, but at the end of the day made a net loss of £40. One of the problems he faced in his latter days was a shortage of labour: the canal age had reached what became known as the mania years and was paying good wages – and it was clear that investors were now much more interested in putting their money into waterways than they were in spending it on roads.

In 1792 he retired to Spofforth, near Harrogate, where he lived with another daughter and her husband. This account of his life ends with few details apart from a description of a visit to his old friend and comrade Colonel Thornton at Christmas 1792. He remained remarkably active in his latter years. In January 1795, he walked to Green Hammerton, covering the 10 miles in a very respectable three and a half hours, and then walked on to Knaresborough. He died at Spofforth on 26 April 1810 at the age of 92. He was buried in Spofforth churchyard, with this epitaph on the headstone – one suspects he might well have written it himself:

Here lies John Metcalf, one whose infant sight
Felt the dark pressure of an endless night;
Yet such the fervour of his dauntless mind,
His limbs full strong, his spirits unconfined,
That, long ere yet life's bolder years began
The sightless efforts mark'd th' aspiring man;
Nor mark'd in vain – high deeds his manhood dared,

And commerce, travel, both his ardour shared.
'Twas his a guide's unerring aid to lend –
O'er trackless wastes to bid new roads extend;
And, when rebellion reared her giant size
'Twas his to burn with patriot enterprise;
For parting wife and babes, one pang to feel,
Then welcome danger for his country's weal.
Reader, like him, exert thy utmost talent given!
Reader, like him, adore the bounteous hand of Heaven.

How important was John Metcalf to the history of engineering? He is generally agreed to be the first professional road builder, although that only occupied a part of an amazingly varied life. It is difficult to be precise, but he was certainly responsible for building about 180 miles of sound, well-constructed turnpike roads. He set a standard that was by no means the case with other road builders, establishing the need for sound foundations and adequate drainage. Whether his better-known successors such as Thomas Telford learned directly from his example or not is uncertain, but he deserves full recognition for his achievements. They are, of course, all the more remarkable given his blindness. He must have been a man of great energy and enthusiasm – as willing to take on the responsibility of a major construction project as he was to accept a bet in the pub. He could walk distances in a day that would tax a fit, sighted man and was always up for a gallop with the hounds. For once that much overused word 'unique' seems wholly applicable.

2

JAMES BRINDLEY

James Brindley was born at Tunstead, part of the parish of Wormhill, a Derbyshire village 4 miles east of Buxton, in 1716. Samuel Smiles always liked his heroes to have a suitably heroic start in life, by providing them with disadvantages to overcome. In this case, the father was portrayed as a wastrel, who frittered away the family income on frivolous pursuits and activities such as hunting and bear baiting. This meant that the boy had to be brought up, more or less, by his mother, with little inheritance to look forward to. There is no evidence to support this, and some to suggest the opposite. The family moved to Leek in Staffordshire in 1726 and three years later were able to purchase a farm nearby. As Boucher points out in his biography of Brindley, the father's will shows that he actually left a considerable legacy since, as well as gifts to his sons, there was to be an annual payment to his widow of £150 5s, equivalent to a very handsome amount worth more than £30,000 today.

Image: James Brindley. (City Museum, Stoke-on-Trent)

The young James Brindley seems to have had a rudimentary education, judging by the written records of his that survive. In notes that he made while working on making a steam engine, he described an 'engon' and, when things were not going well, he recorded 'midlin louk' and later recorded it as being 'at woork'. One can hear in these words his north Midland accent, and suggests that he wrote phonetically. He did, however, according to Smiles, show an interest in mechanical objects, watermills in particular, and he would dam up a stream to make a mill race and install little watermills that he made himself. He would have done work on the farm as soon as he was old enough, but at the age of 17 he abandoned his model mills for the real thing. He began a seven-year apprenticeship with a millwright and wheelwright, Abraham Bennett of Sutton near Macclesfield.

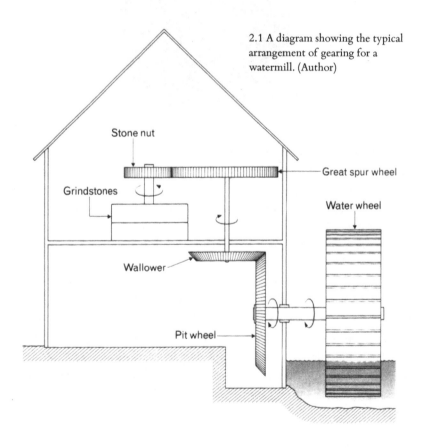

2.1 A diagram showing the typical arrangement of gearing for a watermill. (Author)

Stone nut

Great spur wheel

Grindstones

Water wheel

Wallower

Pit wheel

The work of the millwright and wheelwright was craft based, instruction being passed down through the generations. Even the simplest mill is a relatively complex piece of machinery. In the ordinary water-powered grain mill of Brindley's time for example, the first problem to overcome was how to convert the motion of the water wheel, turning vertically about a horizontal axle, into the movement of mill stones, rotating horizontally. This was achieved through gears. The pit wheel turned on the same axle as the water wheel. It had a bevelled edge that engaged with a similar horizontal wheel, the wallower. Above the wallower on the same shaft was the great spur wheel, and that engaged with a much smaller gear, the stone nut. Because of the difference in size, the stone nut would move much faster than the spur wheel. It was this that turned the grindstones. The main task of the millwright was getting all these different elements perfectly co-ordinated. One can think of the millwrights as being the first skilled mechanical engineers. They also had to understand water management when constructing the weirs and mill streams that ensured there was a good supply of water to power the whole system. It was, or should have been, excellent training for a young man who would eventually make a reputation as a leading engineer.

The task of the wheelwright is as complex, in its way, as that of the millwright. Here the problem is created by the way in which the spokes join the hub to the rim. They are slightly dished, running out at an angle from the hub, and joined by mortice and tenon joints that have to be cut at the same angle, rather than straight. This is a difficult job to get just right, and it was almost young Brindley's downfall. According to Smiles, the master was fonder of the bottle than the workshop, and took little time to teach his apprentice, though he would have been paid to do so. Consequently, the boy was left to find his own way as best he could. He was once left on his own to make a wheel, and got the whole thing the wrong way round: instead of sloping outwards, the spokes sloped in, rendering the whole wheel useless. The master threatened to tear up his indenture papers and send him off to find work as a farm labourer. In the event, the threat was not carried out, and Brindley began paying careful attention to how everything in the workshop was carried out, and in time became a highly proficient craftsman, though still an apprentice.

His first major task outside the Sutton workshop was to go to a small silk mill near Macclesfield, where, rather it seems to the surprise of his master, he finished the job to such a degree of perfection that the mill superintendent publicly announced that Brindley would soon be a better workman than the rest – which no doubt pleased the young man, but would hardly have endeared him to his fellow workers. Inevitably, Smiles has a story showing the brilliance of the young engineer-to-be. Bennett was asked to work on a paper mill, which was not something with which he was at all familiar, so he had to go and see one for himself to see how it worked. He made notes and began working on the new mill, but there was a strong feeling locally that he was making a mess of it. Brindley decided, on his own initiative, to try and resolve the problem. He left work on the Saturday, walked 25 miles to the paper mill, absorbed all the details of the working and walked back in time to start work on Monday morning. This story really stretches belief to breaking point – a 50-mile walk in a weekend, with time in between to take in and understand the workings of a mill! He then proceeded to tell his master where he had got things wrong and the mill was completed successfully thanks to the apprentice's superior judgement.

That Brindley did become a highly competent millwright is not in question, and it seems that in time Bennett felt able to spend more and more time in the inn, leaving Brindley in charge of the workshop. When Bennett died, Brindley carried on for a while until all the work in hand had been completed and the finances had been sorted, and then he left in 1742 to set up in business for himself in Leek. At this stage, he was mostly making wagon wheels and doing repair work on local mills, but he was ambitious and realised that, to be more successful, he needed to have premises where there was a greater demand. He rented another workshop, 9 miles away at Burslem, then at the very heart of a booming pottery industry. The premises were owned by members of the Wedgwood family, relatives of the famous Josiah, who was to feature as a major figure in Brindley's life, but at this time had yet to establish works of his own.

The move helped Brindley prosper, no longer limited to repairing mills, but designing mills and building them, with at one time as many as eight workmen in his employ. Some were conventional grain mills, and one example survives, a corn mill built by the River Churnet in Leek. It is not exactly as built – originally there was a workshop attached to the side. The frontage was originally three bays wide, instead of the present two, and the side now facing the road would originally have been a dividing wall that had then been rebuilt. Among the numerous Wedgwoods engaged in pottery were two brothers, John and Thomas, who were among the first to introduce salt-glazed stoneware.

It was a time of change. The earthenware of the Potteries was made using dark clay and, if a white finish was needed, that had to be supplied by a heavy glaze. But in the eighteenth century it was discovered that by adding powdered flint to the clay, it would come out of the kiln after firing at a much lighter shade. The process began by calcining the flint, heating it in a kiln to make it more brittle and then pounding it to a powder. When the latter process was done with dry flint, the dust flying out caused great harm to the workers, who suffered from lung disease. The answer was to grind the flint in water and then dry it out. The Wedgwood brothers commissioned Brindley to build a flint windmill. The main drive shaft was used to turn four arms, each of which were loaded with hard stones. They swept around inside a large tub filled with the flints and water, gradually reducing the mass to a thick paste. It was not an immediate success: on the very first day of operation, a very high wind got up and simply blew off the windmill's sails. It was a setback, but not a disaster: Brindley simply made the whole good again. He went on to build other flint mills, mostly water powered. A mill of the type built by Brindley still survives at Cheddleton beside the Caldon Canal.

One task that Brindley undertook was to have a profound effect on his later career. John Heathcote owned the Clifton estate in Lancashire and it included a mine that was aptly known as Wet Earth colliery, as it was subject to flooding and had become unworkable. Heathcote's wife had originally lived near Burslem, and it was there that he heard of the ingenious Mr Brindley. The two men looked over the plans of the mine and Brindley declared that he had the solution and was invited to Clifton to visit the site. The mine shaft was close to the River Irwell at a point where it goes through a U-bend. Brindley's plan involved constructing a 30ft diameter overshot waterwheel – one in which the water falls from above the wheel into buckets set around the rim, the weight of which turns the wheel. To get the water to the right height meant going back from the site to a point on the river that would be over 30ft above the site to get the power to turn the wheel. That turned out to be roughly three-quarters of a mile away. At that point, Brindley built a weir, from above which water could be drawn off past the mine. At first the water was carried away in a brick-lined tunnel, until it met the river again. There a pit was dug and the water carried under the river through a siphon, to emerge at the far side, where it was carried in a leat to the top of the wheel pit. The wheel worked pump rods through cranks and connecting rods. It was an ingenious and complex scheme, but it worked and remained in use until 1867 when the water wheel was replaced by a turbine; and that remained in use until the pit finally closed in the twentieth century.

One of Brindley's commissions in 1755 was for work on a new silk mill at Congleton in Cheshire. He was required only to supply the water wheel and associated gearing to the drive shafts. The more detailed work for the actual machinery and the superintendence of the whole works was to be the job of a master joiner and millwright called Johnson. He would issue instructions to Brindley verbally, but never let him see any actual plans of what was needed. It soon appeared that things were not going at all well and, in spite of Johnson's early assurances, it became obvious that nothing worked. The owners turned to Brindley to ask him if he was prepared to finish the work, which he agreed to do, but Johnson still refused to show him the plans or the model of the machinery. Brindley, however, had worked on other silk mills; machinery for throwing silk had first been used at Derby as early as 1721, based on machinery originally designed in Italy. Consequently, he had a good idea of what was needed. According to Smiles, he made his own improvements, including a device for winding the spun silk threads directly onto bobbins. He made his own machines for cutting the tooth-and-pinion wheels for the machine, although there is no record of what form it took. The mill was a success and added yet further to Brindley's reputation as a capable and inventive millwright.

At the start of the eighteenth century, miners were having more and more difficulty in preventing water flooding the pits as they were sunk to ever greater depths. The power available from wind and water was no longer proving sufficient. The solution was found by Thomas Newcomen with a working steam engine. In order to work, pump rods have to go up and down in a steady rhythm. In Newcomen engines, the rods were suspended by chains from an overhead beam, pivoted at its centre. The rods would drop under their own weight: what was needed was some sort of force to pull them upwards. At the opposite end of the beam, a piston was suspended from chains and enclosed within an open-topped cylinder.

Steam was then injected below the piston, then cooled by spraying with water. The steam then condensed, creating a partial vacuum, and air pressure now forced the piston down, lifting the rods at the opposite end of the beam. Once pressure was equalised, gravity again took over, pulling the rods down again and the beam see-sawed to a regular beat and the rods rose and fell.

The machine, usually referred to as an atmospheric engine, since it was air pressure that moved the piston, was first installed at a colliery near Dudley Castle in 1712. It was a success and soon the engines were gently nodding their way at mines across the country. They were, however, extremely inefficient, in that a vast amount of heat was needed to produce a comparatively small amount of work. This was not so much a problem at collieries where coal was readily available, but proved much more serious in other areas, such as among the metal mines of Devon and Cornwall.

Several engineers tackled the problem of increasing efficiency, and James Brindley was one of them. He had been involved in constructing a Newcomen type of engine for Thomas Broad of New Fenton, which as far as we know had no novel features. But when it was at work, he would have been aware of the huge fuel consumption. He recognised that the root cause was the fact that at each stroke the whole cylinder was cooled by the water spray, and had to be heated up again afterwards. He reasoned that heat losses could be reduced if the cylinder could be made of a substance that was far less conductive than the iron commonly used. He chose wood for the next engine, built for Miss Broad, presumably a relative of Thomas's, at Fenton Vivian.

There are detailed notes by Brindley on the progress of this engine. He went to the famous Darby ironworks at Coalbrookdale for the metal parts, such as the cast iron steam pipes. It seems that the cylinder was bound round with hoops and, after a few false starts, he was able to report on 7 April 1758 – 'Engon at woork 3 days' and a little later 'driv a-Heyd'. The triumph was short lived: by 21 April it had stopped working. The sad fact is that it was never going to be workable with a wooden cylinder. For the engine to operate, the heat has to evaporate rapidly to create a vacuum, which was possible with iron, but not with timber. The engine is later reported to have been amended, presumably by replacing the wooden cylinder with a conventional iron one. Brindley did not repeat his attempt to improve on the Newcomen engine in any meaningful way.

Although, there were no more attempts at engines, he did experiment with improving boilers. The original Newcomen boiler was basically just a large pot full of water sat on top of a fire – an overgrown version of a kettle on a stove. His first attempt was to have a wooden boiler with a cast iron firebox inside it. That failed. The next version had the firebox set within a frame of brick or stone. The boiler was fed by gravity. A flap that admitted the water was operated by a float. He was sufficiently confident of this device that he took out a patent in December 1758. It never lived up to expectations: the device failed when the steam pressure rose much above 3 pounds per square inch (psi). Brindley was not destined to make his name as a great mechanical engineer, but he was soon to find the vocation that was to make him famous. He was about to become a canal engineer.

By the middle of the eighteenth century, there had been huge improvements in making Britain's rivers navigable. The old flash locks had almost all been replaced by pound locks – the locks we all know today, chambers closed off by gates at both ends. This involved building a weir to hold back a head of water and then cutting an artificial channel with the lock in it to bypass the obstruction, allowing an orderly move from one level of water to the next. But there was a limit to what could be achieved, and there remained large parts of the country with no easy access to navigable waterways. And, as we have seen in the story of Blind Jack, roads of the period left a lot to be desired. The next step was obvious: to make artificial canals, independent of natural waterways. A start had been made when an Act was passed in 1755 authorising the construction of the Sankey Brook Navigation, the first modern canal and one that was to run from the Mersey to St Helens. The name is misleading, as there was no question of making the brook navigable, but it was used as a feeder to supply the artificial canal with water. It attracted little attention: it had no dramatic features and the name suggested that it was no different from other navigation schemes of the time. The next canal was very different.

The story begins with Francis Egerton, the younger son of the first Duke of Bridgewater. He was born in 1736 and was largely ignored by the family, who seemed to have considered him rather slow witted. All the attention was on his elder brother, who was destined to inherit the title and was given a suitable education to fit him for an aristocratic life. The duke died in 1745, but his successor only lived another three years before he too died, and the title passed to 12-year-old Francis. Now there was a scamper to give him the education that no one had previously thought necessary, which ended in that essential rite of passage for eighteenth-century aristocrats, the Grand Tour of Europe. He was sent off in the care of Richard Wood, a scholar and antiquarian. Their first stop was Paris, where the 17-year-old showed little interest in works of art, but, in Wood's view, far too much interest in young actresses. He was hastily moved on down to Italy, where he was encouraged to purchase antiquities that were crated up and sent back to England – where, it was said, they were never opened.

If he was not especially interested in the ancient world, the boy was fascinated by the modern. On his travels he got to visit the Canal du Midi. This immense waterway cut across France, linking the Mediterranean to the Atlantic. Built between 1666 and 1681, it contained remarkable features: locks grouped together into staircases, aqueducts and a tunnel. Nothing on that scale had yet been attempted in Britain. It impressed the young man, but on his return to Britain he began to live the life of aristocratic society and he fell in love with the beautiful young Duchess of Hamilton. He was, however, not really fitted for the life, being somewhat puritanical and strait-laced. When he heard that the duchess's sister was involved in scandal, he instructed her never to speak to her again. The duchess, very reasonably, refused, and the affair was at an end. The young duke swore never to have anything to do with women again – it was said he even refused to have female servants in his house. At the age of just 22 he retired to his estates near Manchester.

Having abandoned frivolity, he now turned to business and began to look for ways to increase the profits from his coal mines at Worsley, near Manchester. The answer was eventually to be a scheme that involved firstly a technique common to many mines: constructing a sough that would act as a drain to remove water from the mines. Where this differed from most similar schemes was the idea that the sough should be built wide enough to take boats that could be loaded at the coal face and then brought out of the mine. The second part of the scheme that was entirely new was to continue from the mine entrance with a canal that would carry the coal on into the heart of Manchester. Here we come into controversy. Smiles gives virtually all the credit for the scheme to Brindley, who, according to his version, was not simply the mastermind behind the whole idea, but the engineer who planned and supervised all the works.

The most reliable source of information would appear to be an article by John Farey, a distinguished engineer in his own right, who wrote a biography of the duke's agent, John Gilbert, in Rees's *Cyclopaedia* published in 1819. In this account, the duke and Gilbert got together to discuss the idea of the underground and overground canals. How far the discussion was influenced by the duke's recollections of the Canal du Midi and how much was due to Gilbert is uncertain. However, they were sure enough of their plans to apply for an Act of Parliament for a canal that would run from the mines at Worsley, staying to the north of the River Irwell and ending at Salford, remaining on the level and lock free throughout its length. The Act was passed in 1759 and work got under way on the underground system under Gilbert's direction. According to Farey: 'The tunnel was entirely executed as planned by Mr Gilbert; who, being acquainted with Mr Brindley as a neighbour, and knowing him to be a very ingenious and excellent mill-wright, engaged his assistance in the conduct and completion of this arduous undertaking, and introduced him to the Duke for this purpose.' This unequivocally puts Gilbert in the more important role, though he no doubt drew heavily on Brindley's experience in draining the Clifton mine.

2.2 The Barton aqueduct carrying the Bridgewater Canal over the River Irwell.

The original idea had been for the coals to continue their journey from Salford to Manchester, a town that had doubled in size in just forty years and where the demand for coal was high. However, the owners of the Mersey and Irwell Navigation Company were unwilling to co-operate without substantial payments. It was now that the duke took a bold decision. If he could not join the Irwell, he would leap over it. He put a new proposal to Parliament, this time for a canal that would go right into Manchester and cross the Irwell on an aqueduct. This was a bold decision, for although the duke had seen aqueducts in France, neither he nor Gilbert and Brindley had any experience in building such structures. He was also taking on powerful interests who would oppose him, and that would involve legal costs. Also, because of the novelty of the whole plan, it was impossible to estimate with any accuracy just how much the whole scheme would cost. The duke was a wealthy man, but his purse was not bottomless.

Once the Bill was before Parliament, the Mersey & Irwell tried to placate the Duke by offering to reduce their exorbitant tonnage rates, and when that belated offer was rejected they began issuing pamphlets and petitions, claiming that the new canal would be their ruin. In fact, they had for some time been gaining themselves a bad reputation for high fees, lack of maintenance of the waterway and delays to traffic, which made their claim to being innocent victims more than a little dubious. The Bridgewater group replied with a pamphlet of their own, decrying the petition that was being touted for signatures:

> Permit me then to remind you to be on your guard, consider coolly what you are about, think of men and times past, examine whence this petition comes recommended, and you'll find it to be at the appointment (perhaps the modest request, or rather, the awful and powerful sanction) of the truly honourable body of Manchester Navigators.
>
> I might have spared the compliment, for trees are known by their fruits, and none can be mistaken in their judgements of the uprightness of these gentlemen's intentions: their past betrays them, the facts are recent, witness their attempts when Sankey-wharf was first erected, their stoppage of vessels and goods on the most base and frivolous pretences, their long and extravagant freights, tonnage, wharfage etc.

The duke won the battle, and the Act was passed on 24 March 1760. Construction of the canal could begin. As work progressed, he became more and more pressed for money. He sold off his London house – he had no intention of ever rejoining Society. Most projects of this kind would have been financed by selling shares and if, which was often the case, cost estimates proved to be wildly optimistic, the shareholders could be called on for further investment or new shares could be issued. But in this case the whole project was being privately financed. He raised money on his estates and borrowed from relatives, including Lord Gower, who was an enthusiastic supporter of canals, from Manchester businessmen who stood to benefit from the enterprise, and from the bank of Child and Co. At times things became so desperate that when the local vicar called round hoping to get repayment of a debt, the young duke was not to be found – he was hiding in a hay loft. They were desperate times and, on this occasion, we have something as near as we can get to a first-hand account. In his autobiography of 1875, Sir John Rennie described a meeting between his brother and a Mr Bradshaw, who was to become manager of the Bridgewater Canal:

> Pointing to a little whitewashed house, near the Moss, about half a mile distant, he said to my brother: 'Do you see that house? Many a time did the late Duke of Bridgewater, Brindley and myself spend our evenings there during the construction of the canal, after the day's labours were over; and one evening in particular we had a very doleful meeting. The Duke had spent all his money, had exhausted his credit, and did not know where to get more, and the canal was not finished. We were all three in a very melancholy mood, smoking our pipes and drinking ale, for we had not the means to do more, and were very silent. At last, the Duke said: "Well, Mr Brindley, what is to be done now?" Brindley said: "Well, Duke, I don't know: but of this I feel confident as ever: if we could only finish the canal, it would pay very well, and soon bring back all your Grace's money."'

Although written more than a century after the event described, it provides an interesting insight into the relationship between the duke and Brindley, and it is notable that there is no mention of Gilbert being present.

The major structure on the canal was, of course, the aqueduct that would carry it across the River Irwell. Peter Lead, in his biography of

John and Thomas Gilbert, *Agents of Revolution*, states categorically that Gilbert designed it and all Brindley had to do was follow his instructions and build it. There does not seem to be any firm evidence to support this view, but equally there is nothing to confirm that Brindley designed it either. It was very much a collaborative venture, and one could take the middle line and say that perhaps both men had a part in the design. Whoever, it was, they had to face a certain amount of derisive comment, sometimes from those who should have known better. One anonymous engineer is said to have remarked that, though he had often heard of castles in the air, this was the first time he had actually come across someone proposing to build one.

The basic structure is not very different from a bridge, only differing in that there is a trough full of water on top instead of a roadway. One particular difficulty lay in making sure it was watertight. The same problem applies to the whole canal. One cannot simply dig a wide, deep trench and fill it with water or it would seep away through the ground. The answer was to puddle it, mix a gooey concoction of clay and water, and use it as an impermeable lining: Brindley would, of course, have been familiar with this from his days as a millwright constructing leats. In the case of the Barton aqueduct, however, a problem appeared as soon as the water was let in, when one of the arches seemed to be giving way. When confronted with a problem, Brindley tended to take himself off to a darkened room and lie down, with no distractions, to think of a solution. In this case, it turned out that he had been too cautious and laid on such a heavy load of puddled clay that it was causing the strain. The water was drained away, much of the puddle removed, and the arches stood firm.

The canal ended at Castlefield, where coals for Manchester were unloaded at the foot of a steep hill. It was close to the River Medlock, a tributary of the Irwell, and the river water was used to power two water wheels. One of these was set beside a shaft, leading down from the top of the hill to the canal. Boats floated down a short tunnel to the foot of the shaft, where the wheel provided the power for a hoist to lift cargo up the shaft. The second wheel was used in a canal warehouse owned by Gilbert and his partner John Henshall.

The canal was a great success, and Barton aqueduct became a tourist attraction. One anonymous writer was quite overcome by the sight of the Duke's canal:

His projector, the ingenious Mr Brindley, has indeed made such improvements in this way, as are truly astonishing. At Barton bridge he has erected a navigable canal in the air; for it is as high as the tops of trees. Whilst I was surveying it with a mixture of wonder and delight, four barges passed me in the space of about three minutes, two of them being chained together, and dragged by two horses, who went on the terras of the canal, whereon, I must own, I durst hardly venture to walk, as I almost trembled to behold the large River Irwell underneath me, across which this navigation is carried by a bridge, which contains upon it the canal of water, with the barges in it, drawn by horses, which walk upon the battlements of this extraordinary bridge.

What few commentators mentioned or even knew about was the extensive labyrinth of narrow canals inside the hill at Worsley, stretching for many miles. The system was served by special boats, just 4ft wide, known as 'starvationers' because of their slender lines and prominent ribs. These were the barges, seen chained together, that our anonymous tourist would have seen. It was not, however, the engineering features of the canal that attracted the attention of merchants and industrialists, but the fact that, when it opened, the price of coal in Manchester was halved.

In 1762, the Duke obtained an Act authorising an extension of the canal to the Mersey at Runcorn. Here, Brindley had new challenges to face. In general, he preferred to keep a canal on one level, even if it meant extravagant curves to avoid obstacles. However, there were two sections he was unable to get round – the valley of the River Bollin and the low-lying, swampy land of Sale Moor. He had no alternative other than to cross them on high embankments. A letter of 1 July 1765 in the *St James's Chronicle* described the situation at the latter:

He has finished the cut quite across Sale Moor, and will soon compleat it over the meadows on each side of the River Mersey; the entrance to which from the low and boggy situation was, by men of common understanding, deemed to be *ne plus ultra*. At this place, Mr Brindley caused trenches to be made, and placed deal planks in an erect position, backing and supporting them on the outside with other balks laid in rows, and screwed fast together; and on the front side he threw the earth and clay, in order to form the navigable canal. After thus finishing forty yards of his artificial river, he removed the balks, and placed them again where the canal was designed to advance.

Even Brindley could not remain on a level for ever, for the canal eventually arrived at a point high above the Mersey. To close the gaps, Brindley constructed a flight of ten locks, arranged in groups of two. In these pairs, a boat going down would pass straight from the top lock into the lower one. This was a broad canal, and to succeed it needed to be able to take the sailing barges already in use on the river, the Mersey flats that were normally just over 50ft long and 14ft broad. The locks each had a fall of 7ft apart from the bottom lock, which was 22½ft so that it could be used at all states of the tidal river. The flight was opened with great ceremony on 1 January 1773 and the 600 workmen employed were feasted 'with an ox roasted whole and plenty of liquor'. The one man who was not there was James Brindley. He had died in September the previous year. But almost from the start of construction he was already engaged in a wide range of other canal projects. He did at least live long enough to see one of his prophecies fulfilled; the success of the Bridgewater Canal meant that the duke did indeed recover his fortune.

Whatever the merits of John Gilbert might have been and whatever his share in planning the works, to the public at large Brindley's was the name that was known. When a new canal was being planned then he was the man they wanted to be in charge. John Farey, when writing about Gilbert's role, decided that he 'was probably so modest and unassuming, that he did not during his life-time lay claim to the honour that was due to him'. In any case, he was extremely well situated in working for the Duke of Bridgewater and content to stay in his employment.

Two new canals were approved by Parliament in 1766, the Trent & Mersey and the Staffordshire & Worcestershire, both of which were to use Brindley as their chief engineer. The Trent & Mersey, also known as the Grand Trunk, was a canal built on a scale never seen before in Britain. One of the leading proponents was Josiah Wedgwood, who had already established himself as an important and innovative manufacturer. He had developed a light-coloured earthenware pottery, known as creamware. This became popular and even more fashionable when it was discovered that a dinner service had been ordered by Queen Caroline in 1765 – at which point all the pottery of that type was known as Queensware.

Wedgwood was obviously a good PR man as well as an excellent potter. But for its success, creamware needed two ingredients: light-coloured clay from Devon and Cornwall and flints from East Anglia. He was aware that Brindley had already carried out an initial survey for Lord Gower for a possible canal linking the Trent and the Mersey, and he realised that this would be a huge advantage to his works. Instead of having to bring clay and flints overland, the clay could be brought round the coast to the Mersey and flints brought up the Trent, and then they could come the rest of the way by water. He began getting information about canals at a meeting on 11 March 1765:

> On Friday last I dined with Mr Brindley, the Duke of Bridgewater's engineer, after which we had a meeting at the Leopard on the subject of a Navigation from Hull, or Wilden Ferry, to Burslem agreeable to a survey plan before taken. Our Gentlemen seem very warm in setting this matter on foot again, and I could scarcely withstand the pressing solicitations I had from all present to undertake a journey or two for that purpose.

Wedgwood went first to a friend, Thomas Bentley, of Liverpool, who was later to become his partner and got him involved in the scheme. On his travels he also visited Lichfield, where he met one of the most extraordinary men of the time, Dr Erasmus Darwin. As well as being a doctor of medicine, he was an amateur scientist and poet. The Darwin and Wedgwood families were always very friendly and Darwin's son married Wedgwood's daughter. In one of his long narrative poems he put forward an idea of the evolution of species – an idea that his famous grandson Charles fleshed out by demonstrating a mechanism through which species developed and changed.

So, Wedgwood was very used to mixing in very intellectual company – he was a regular attender at the meetings of the Lunar Society of Birmingham, where discussions centred on all the latest ideas, especially in the sciences. He moved in a very different circle of friends from those associated with Brindley. Yet, in spite of this, the two men got on extremely well and Wedgwood often spoke of how much he admired the engineer, on one occasion noting that he always enjoyed his conversation and never failed to learn something new. In other ways, they were very different. Wedgwood's work was inevitably involved as much in the arts as in the sciences and the practical life of a manufacturer. The arts were of no interest to Brindley. As an engineer involved in planning canals, he would sometimes have to go to London to be quizzed by the Parliamentary committee working on an enabling Act. It was on one of these occasions that he was taken to the theatre to see the famous actor David Garrick in *Richard III*. He found the whole experience terribly upsetting, was unable to get any work done the next day and swore never to go a theatre again.

Preparations got under way for applying for an Act of Parliament to authorise construction. This canal, unlike the Bridgewater, was to be financed by the formation of a public company and the sale of shares. There was the usual opposition from other parties, who swore the arrival of a canal would ruin their own businesses and leave them destitute, and pamphlets were put forward opposing the scheme – one from 'The Gentleman and Tradesmen of Warrington' argued that the canal 'could only be desired for *private* views, or to make a *more lucrative* Job for Engineers'. Wedgwood persuaded Bentley to write a pamphlet explaining the benefits the canal would bring and asked Darwin to help. The two authors quarrelled over the minutiae of the wording to Wedgwood's extreme annoyance: 'Must the Uniting of Seas & distant countries depend upon the choice of a phrase or a monosyllable? Away with such hypercriticism, & let the press go on, a Pamphlet we must have, or our design will be defeated, so make the best of the present, & correct, refine, & sublimate, if you please, in the next edition.' They were successful and Wedgwood made a note of the principal characters who would be responsible for seeing the work through, with a wry note on his own position:

James Brindley Surveyor General £200 per ann.
Hugh Henshall Clerk of the Works £150 per ann. For self & clerk
T. Sparrow Clerk to the Proprietors £100 per ann.
Jos. Wedgwood Treasurer at £000 per ann. Out of which he bears his own expenses

Meanwhile, Brindley was setting about the serious business of selecting the best route. He began by riding along the possible route to get a feel of the land, before the essential work of carrying out an accurate survey could begin. The eventual route chosen was to run from an extension of the Bridgewater Canal at Preston Brook to join the Trent near

Nottingham. It would have been logical to make the new canal accessible to the barges of the two rivers, as he had done for the Mersey flats at Runcorn. However, he was faced with an obstacle that he simply could not go round, the hill at Harecastle, between Tunstall and Kidsgrove. The only solution was a tunnel that would need to be around a mile and a half long. He had some experience of tunnelling with his work at the Clifton colliery, but nothing on this scale. At that time the only canal tunnels in Britain were the narrow waterways in the Worsley mines. The idea of constructing a tunnel that would take vessels over 14ft wide was simply too daunting. If he halved the width of the tunnel, he would do more than halve the work required. If you think of the opening as being circular, which it isn't exactly, but is near enough for this purpose, then the area of the opening is given by the square of the radius multiplied by the constant pi. Again, with very rough figures, to take a wide boat the opening would be roughly 160sq ft, but if he halved the width, it would be just 40sq ft. So that is what he decided to do, and having taken that

2.3 The two tunnels at Harecastle Hill on the Trent & Mersey Canal: the original Brindley tunnel is on the right. (From Smiles, *Lives of the Engineers*, 1862)

decision, he then reasoned that if only vessels 7ft wide could pass through the tunnel, there was no point in making locks to take wider vessels. And that decision meant that he carried the same dimensions to other canals in the region that had connections with the Thames & Mersey, and he gave the canal system its best-known vessel – the narrowboat.

Construction of a tunnel is a comparatively complex affair. The first essential is to get the profile of the hill, using standard surveying techniques. In his portrait (p. 27), Brindley is shown with one arm resting on a theodolite, which he would have needed for the survey. The instrument basically consists of a telescope mounted on a tripod. There is a levelling device, and the telescope can be moved both vertically and horizontally. To measure the slope of a hill, the surveyor's assistant would take a long, calibrated pole for a fixed distance from the point where the theodolite was set up. He would use a chain to measure that distance – it was a practical device, as it could be rolled up when not in use and, being metal, was not subject to stretching and distortion. The chain was a standard length of 22 yards – which, incidentally, is how that became the distance between wickets on a cricket pitch. The theodolite would be set along the line the tunnel was to be taken and, with the telescope set level and pointing in the precise direction needed, a sight would be taken of the pole. The point on the pole corresponding to the level of the telescope gives the height above the starting point. This would then be repeated, moving forward a chain at a time, until a point was reached on the far side of the hill where the height was equal to that at the starting point. It is then a simple exercise in trigonometry to create a profile of the hill. Once the profile has been established, the line of the canal can be marked out and, knowing the height above the original level at any point, shafts can be dug down to that level, in this case fifteen in all. Workmen can then get started, using the known compass bearing to work out from the bottom of each shaft in both directions and working inwards from the two ends. If everything has been done with precision, they should all meet in a straight line. Brindley's tunnel has long been disused, but a later survey found it to be distinctly crooked.

One of the problems Brindley faced was that he had no idea what sort of material he might find deep into the hillside. In fact, it was very mixed. In some places the tunnellers met hard rock that had to be blasted away. This involved drilling holes by hand, then filling them with gunpowder and adding a slow-burning fuse – and for the workers involved getting as far away as quickly as possible. In other spots they met soft, wet ground like quicksand. Pumps had to be installed to keep the workings dry, at first worked by wind or watermills and, when these proved inadequate, a Newcomen-type engine had to be installed on the site to take over the task. A stove was used to suck out the methane that was produced where they met coal measures. There was an early air of optimism, and one visitor to the site recorded that Brindley 'handles rocks as easily as you would plumb-pies, and makes the four elements subservient to his will'. The elements were not to be that easily subserved. The engineer was confident that the whole project would be finished in five years and, when one of the committee doubted that was possible, Brindley took a wager of £200. He would have lost, for the whole work took eleven years, but Brindley never had to pay out, for he had died before it was completed.

2.4 The Trent & Mersey passing Josiah Wedgwood's factory at Etruria. (Josiah Wedgwood & Sons Ltd)

The canal was completed in 1777 and Wedgwood, who had fought so hard for its cause, had the pleasure of being able to open his new and efficient pottery on its bank at Etruria. On the opposite side of the canal, he built a splendid new house for himself and his family, Etruria Hall. He had hoped to persuade Brindley to provide him with a graceful, sweeping curve where the canal passed the end of his grounds, but all he got was an uncompromisingly straight line.

The other canal on which Brindley was working at the same time did not present quite the same problem. It was to link the Trent & Mersey to the River Severn. One question that needed to be decided was where the junction with the river was to be made. One account has it that Brindley's first choice was the town of Bewdley, which already had facilities for dealing with vessels on the river. However, the citizens of Bewdley told the engineer that they had no need for his 'stinking ditch'. The story appeared in *A History of Worcestershire Agriculture*, published in 1939. Further research shows, however, that exactly the opposite was true. The citizens asked Brindley to make the juncture there, but he surveyed the ground and found too many obstacles in his way. Instead, he preferred to take the far easier route following the line of the little River Stour. The new terminus had a basin and was approached from the river by both wide and narrow locks, so that it could be used by both barges and narrowboats for transhipment. In time a fine warehouse and a hotel were built beside the basin and a new town of Stourport grew up around them. Something similar happened at Shardlow at the Trent end of the Trent & Mersey, but on a somewhat smaller scale. One other major feature on the canal was an aqueduct over the River Sow. As the river was not navigable, there was no need to build on such a grand scale, and it is carried over the river on four low arches, but it is approached via awkward right-angled turns.

Ultimately, the idea was to join these two canals to others to form a cross that would join the country's most important rivers: Thames, Severn, Trent and Mersey. The Acts for the Birmingham and the Coventry Canal were passed in 1768 and the Act for the Oxford Canal to complete the scheme the following year. All of them appointed Brindley as their chief engineer. This was an enormous burden for one man to bear, and each company expected him to make their project his top priority. The Oxford Committee tried to keep their engineer under control and passed a resolution that he had 'in no degree complied with the Orders of the Committee', and in the following

2.5 Stourport, where the Staffs & Worcester Canal joins the River Severn. (British Waterways)

year, 1769, they must have complained again about his lack of attendance, because Brindley took offence and resigned. The Committee at once gave way, apologised profusely and expressed their complete confidence in his work. He withdrew the resignation.

In working on these canals, he followed his usual practice of keeping as far as possible to the contours. This can be seen in extravagant form on the Oxford Canal at Wormleighton, north of Banbury, where the canal practically encircles a small hill. It is a curious experience when boating this section to look across and see a boat pointing the opposite way to yours, but actually travelling in the same direction. The northern section of the canal was later improved and straightened, but the old meandering course can still be traced. The original tunnel entrance is next to the church at Newbold and nearby is a canal bridge stranded in the middle of a field. The Birmingham Canal was equally wayward and it too was improved and straightened. In this case, however, the original canal was still very much in use and valued by the new factories and warehouses that had been built along its route. So the curves were left as loops off the main line. One of these, the Soho loop, was home to the famous engineering works of Boulton & Watt.

On two occasions, artificial canals gave way to natural rivers. On the Trent & Mersey, the canal crosses the river on the level, so that all that is needed is a simple wooden bridge across the river to carry the towpath. The Oxford Canal briefly joins the River Cherwell above Oxford. The approach to the river section is via Shipton Weir lock, which unlike most locks is hexagonal rather than rectangular. There is a good reason for this. In general, locks are more or less the same depth, so that the same amount of water is passed down to the next lock in order for it to be filled without draining the pound, the section between locks. Here, the difference in level between river and canal is slight, so the extra area keeps the volume of water the same as a deep lock would have done.

One of the most ambitious of all the canals with which Brindley was engaged is the Chesterfield Canal, which has two distinct sections. The first section from West Stockwith on the Trent to Retford is a broad canal with six locks able to take river barges. From there it is the familiar Brindley narrow canal, with forty locks in 25½ miles to lift it to the summit level. Of these locks, seven are three-rise staircases and two are doubles. Having reached the summit, the next obstacle was the rising ground that was to be pierced by the 2,584-yard Norwood tunnel. After that there was a more gentle descent via fourteen locks to the Rother Valley and then a climb out again by another five locks to Chesterfield. It was an impressive piece of engineering, but was destined not to be completed in Brindley's lifetime, finally being opened in 1776. The Droitwich Canal was one of the engineer's lesser works, a mere 6 miles with eight locks, linking the town to the Severn. It was approved by an Act of 1769 and completed just two years later.

Brindley was constantly in demand – it seemed that everyone who had any sort of canal scheme needed his seal of approval. Sometimes he carried out a survey, made plans and then left it to others who were to supervise and carry out the actual construction. This was the case with the Chester Canal, which began with an ambition to join the Trent & Mersey, but never actually got there. The Bradford and Huddersfield Broad canals that he also surveyed, but in which he took no further part, were rather more successful. Other projects never got any further than the planning stage – including one from Stockton to Darlington, a link that would only be made in the railway age, and one from Exeter to the Bristol Channel. Wedgwood worried about his friend and the vast amount of work he was undertaking, seemingly unable to tun any offer down. In 1767 he wrote to Thomas Bentley:

> I am afraid he will do too much, & leave us before his vast designs are exe-
> cuted, he is so incessantly harassed on every side, that he hath no rest, either
> for his mind, or Body, & will not be prevailed upon to take proper care of
> his health ... I will think Mr Brindley – *The Great, the fortunate money-getting*
> Brindley an object of Pity! & a real sufferer for the good of the Public. He
> may get a few thousands, but what does he give in exchange? His *Health* &
> I fear his *Life* too, unless he grows wiser, & takes the advice of his friends
> before it is too late.

Throughout his working life, he had a constant business relationship with the surveyor John Henshall, who had an attractive daughter, Anne. Brindley first met her in 1762 when she was still at school. Whenever he visited the family, he always had a pocketful of gingerbread for the girl. When she was just 19 years old, he proposed to her and they were married. It seems to us an unlikely match between the teenager and the middle-aged man, but such matches do occur and novels of the period suggest that they were not thought at all odd in the eighteenth century. They settled down in an old but comfortable house at Turnhurst and seem to have had a very happy marriage, during which they had two daughters. When her husband died, Anne remarried and had seven more children.

Brindley's death came in 1772. He had returned home at the beginning of September 1772 and Wedgwood wrote to Bentley to give him the news:

> I told you in my last letter that Mr Brindley was extremely ill, and I have
> the grief to tell you that he is no more. He died the 27th inst. about 12 at
> Noon, and died in a sound sleep, for about 3 o'clock in the morning, after
> giving him something to wet his mouth, he said – *'tis enough – I shall need no
> more,* and shut his eyes, never more to open: he continued to the time of his
> death (about 9 hours) seemingly in a fine sleep, and yielded up his breath at
> last without a single groan.
>
> He has left two young Children behind him, and poor Mrs. Brindley
> inconsolable for the loss of a sensible friend and affectionate Husband.
> What the public has lost can only be conceived by those who knew his
> Character and Talents to which this Age and Country are indebted for

works that will be the most lasting monuments to his Fame and shew to future Ages how much good may be done by one single Genius, when happily employed upon works beneficial to Mankind.

Brindley had dominated the early years of canal construction and he set the pattern of development for the whole of the Midlands of England with his narrow, winding waterways. Paradoxically, the drawbacks to these early canals, that they were too meandering and that they were only usable by narrowboats, which restricted their commercial development in later times, are precisely the qualities that make them so popular with the pleasure boaters of our own age. But even while Brindley was still alive there was a new generation coming along with different ideas, who were not necessarily daunted by his reputation and were quite willing to challenge him. One canal that engaged Brindley to carry out initial surveys was the Forth & Clyde, but he was not the only engineer consulted. The other involved wrote this scathing letter:

Mr Brindley recommends to begin at the point of partition, because, he says, it is his 'constant' practice to do so, and, in the present undertaking, it seems particularly advisable 'on many accounts'; but pray, Mr Brindley, is there no way to do a thing right but the way you do? I wish you had been a little more explicit on the many accounts: I think you only mention one, and that is to give more time to examine the two ends: but pray, Mr Brindley, if you were in a hurry, and the weather happened to be bad, so that you could not satisfy yourself concerning them, are the works to be immediately stopped when you blow the whistle, till you come again, and make a more mature examination? ... a late author has very solidly demonstrated that every man, how great soever his genius, has a certain hobby horse that he likes to ride; a large aqueduct over a large river does not happen to be mine.

The writer of those words was John Smeaton, who is the next engineer whose life we shall be looking at.

3

JOHN SMEATON

John Smeaton had a very different background from the men we have looked at so far, as perhaps might seem obvious from the language he used in the letter to Brindley quoted on p. 51. His father, William Smeaton, was an attorney in Leeds, and the family had a rather grand home, Austhorpe Lodge, some 5 miles away. It was here that John was born in 1722 and where he was educated until the age of 10, when he went to Leeds Grammar School; he remained there until he was 16. It seems that from an early age he showed a great interest in anything mechanical, and after leaving school he had his own workshop. This might have been no more than a hobby had he not had the great good fortune to become friends with Henry Hindley of York in 1741. Hindley was then 40 years old and already established as one of the country's leading manufacturers of clocks, watches and scientific instruments. He had several inventions to his credit, including a screw-cutting lathe, one of which Smeaton was able to install in his own private workshop.

Image: John Smeaton. (Stephen C. Dickson)

At this time, the parents were beginning to get seriously worried about their son's future: Hindley may have been a brilliant technician and innovator, but he was in trade and William Smeaton was in a profession. So, at 19 young Smeaton was sent off to London to begin studying for the law. We have no means of knowing how much effort he put into mastering the legal profession, but clearly he had no love for the occupation. He did, however, make friends with another reluctant law student, Benjamin Wilson, whose great love was for studying science and had a particular interest in electricity. Wilson knew many of the leading scientists of the day, including members of the prestigious Royal Society, and introduced Smeaton to them. Soon after this both young men gave up law forever: Wilson to pursue his studies in electricity and also to become a very successful portrait painter; Smeaton to return to Yorkshire, where he began making scientific instruments to help Wilson with his work and for making experiments of his own. They shared a common interest in astronomy and soon Smeaton was adding lens grinding for telescopes to his range of skills.

There was one other distraction in his life: he fell in love, with a Miss Banks. Once again, class distinctions rose against him: her family were in trade, and the Smeatons threatened to disinherit him if he continued with the match. Eventually, they gave way slightly, but only on condition that he left Yorkshire for London and set up house there. At this point the girl's father objected to her going to the capital, and the whole affair fizzled out. But by now, any idea of John Smeaton taking up the legal profession was abandoned and in 1748 he set up in business as a manufacturer of scientific instruments in London.

He was successful in his new career and among the instruments he made was an improved compass for use at sea; he also designed a ship's log – an instrument based on a brass spinner and counter to measure a ship's speed. The latter was less successful. In the early 1750s he also began a series of scientific experiments, for which he designed his own apparatus. The most famous were set up to determine which form of water wheel was the most effective. There were four basic types of water wheel. The simplest was the Norse wheel, in which a rotating blade, set at the bottom of a vertical shaft, was set directly into the flowing stream. It was generally known that this was the least efficient, and was rarely used by this time. The others were the undershot, where the paddles on the lower part of the wheel mounted on a horizontal shaft dipped into the mill stream, forcing the wheel to turn; the overshot, described in the previous chapter; and the breast shot where, as the name suggests, the water arrived part way up the wheel. There was considerable disagreement among millwrights as to which was the most efficient, but no one had ever carried out accurate experiments to settle the matter. Smeaton built a model, in which he used hand-pumped water to provide a constant flow and then used that to turn

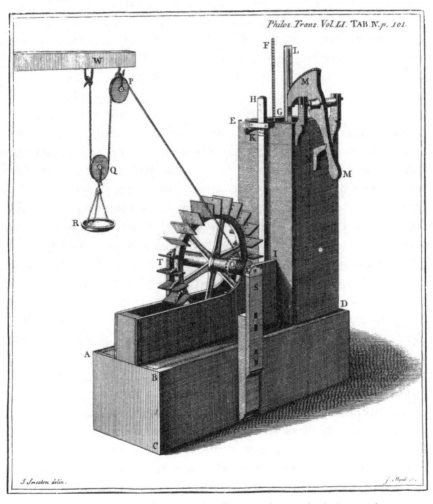

Philos. Trans. Vol. LI. TAB. IV. p. 101.

3.1 Smeaton's experimental rig for testing the efficiency of waterwheels. (From *The Miscellaneous Papers of John Smeaton*, 1814)

the wheels. He measured the effectiveness by how far a weight moved by the wheel via a pulley system could be lifted in one minute. The results were decisive: the overshot wheel was roughly twice as efficient as the undershot. He did not test a breast shot, but its efficiency is, not surprisingly, somewhere between the two.

Perhaps, the news of his interest in mills had spread, otherwise it is difficult to see why, in 1753, he was invited to design a watermill at Halton in Lancashire when he had no previous experience as a millwright. It was obviously a success, for it was to be only the first of many mills he was to design over his career. The following year there was another watermill and a windmill, this time at Wakefield. Scientific instrument making was becoming less and less important. Nowhere in the world was watermill and windmill design more advanced than in the Netherlands where, as well as the more common uses, such as grinding grain, they played a vital role in land drainage. In 1755, Smeaton made the trip across the Channel to see these works for himself and to look at other aspects of civil engineering, including canal and river locks. On his return, he made his way up to Scotland, where he took part in a survey of the Clyde. It is difficult to imagine now, but at the time it was actually possible for a rider to cross the river on horseback in the centre of Glasgow. He was enjoying a successful career, but had hardly produced any work to distinguish him from others in his new profession. That was about to change thanks to a tragedy.

The need for a warning light on the Eddystone rocks off the coast near Plymouth had been known for centuries. Henry Winstanley's lighthouse of 1698 was an exotic but flimsy wooden structure that was blown away in a great gale of 1703. Winstanley himself and the keepers who were there at the time all died. The replacement designed by John Rudyard was altogether sturdier, anchored to the rock by iron bolts, but again with timber cladding. In January 1756 a fire started at the top of the tower, probably caused by sparks from the lantern. The keepers tried to dowse it, but it spread rapidly and soon the whole structure was alight. The men scrambled onto the rocks, where they were picked up by boat and brought safely back to land. The lighthouse, however, had gone beyond repair.

At this period, lighthouses were all built by private companies, who received revenue from charges made on shipping in the Channel. Robert Weston, a major shareholder in the lighthouse company, wrote to Lord Macclesfield, President of the Royal Society, asking him to recommend a suitable man to take on the task of designing a new structure. He recommended approaching Smeaton as someone wholly reliable: 'when the business was stated to him, he would not undertake it, unless he clearly saw himself capable of performing it.' This was a remarkable recommendation, given that it was only three years since he had given up the instrument-making business to concentrate on engineering. He did, however, have the great advantage of being well known to the Royal Society, having attended with his friend Wilson (he was to be elected a Fellow in 1762). He was approached and did take on the task in January 1756.

The difficulties he faced were immense. The rocks lie 14 miles south of Plymouth, only a small part of the reef being normally above the surface, and even that could be swept with water at high tides and when gales blew down the Channel. Just getting men and material to the rocks was itself a major challenge. It was made even more hazardous by the fact that Britain was at war with France and boats could be captured by enemy privateers, or the sailors could be forced into the Navy by press gangs. There would be long periods when nothing could be done and no boats could reach the site, so it was essential that Smeaton made the best possible use of the times when work could be carried out. Good planning was essential, and Smeaton was fortunate that he had the full confidence of Weston, who trusted him to do the job without any interference.

Smeaton studied the Rudyard design, but decided against following his example of using extensive amounts of timber. His lighthouse would be stone. His first problem was to devise how to build on the sloping rocks, and he used a wooden model to try out his ideas. Having concluded that it was feasible to build a stone tower, he then had to decide on its shape. To secure a firm base, it would need to be wider at the bottom than at the top: the Rudyard building had been conical. Smeaton, however, chose to have the sides curve inward from the base, rather like the trunk of a tree. His next problem was how to hold the masonry together: on dry land, mortar alone could have been used, but this was not practical for the lower courses on a site constantly being washed by the waves. Another alternative was to use iron clamps, but the logistics of moving so much metal out to the rocks made that idea equally unattractive. His solution was to have the blocks shaped so that they fitted together with the equivalent of dovetail joints: the solid base was not unlike a three-dimensional jigsaw puzzle.

With the plan for a stone lighthouse approved, Smeaton visited Plymouth in the spring of 1756 and found the ideal man to supervise the actual construction. Josias Jessop was a foreman shipwright, but he had also been in charge of keeping the Rudyard lighthouse under repair, so he knew the rocks well. It was essential for Smeaton to get an accurate survey of the rocks and their surface, so that he could plan the foundations of the tower and how they would be keyed into the surface. He and Jessop set out ten times to sail to the rocks, but only managed to land on four occasions. However, this at least gave Smeaton a good opportunity to assess the difficulties the workers would face. He also had time to think about the stone he would use. Portland stone was an obvious choice, as it is famously good to work, but it is also liable to erosion by the sea. Smeaton's solution was to use this stone at the core of the building, but for the exterior exposed to the elements he would use the hard granite from Cornwall.

Plate XVIII.

THE EDDYSTONE LIGHT-HOUSE.

3.2 The Eddystone lighthouse. (Royal Museum, Greenwich)

Smeaton returned to London, where he made a three-dimensional model of the rock on which the lighthouse was based and a second model showing how the rock was to be cut. The first six courses of stone were to form the solid base, above which the construction changed to blocks arranged in rings. The entrance was set above the fifteenth course and, by the time the twenty-fourth course was reached, the structure was a simpler circular wall. Once work got under way, there was a great deal of organisation needed. A working yard was established near Plymouth Hoe, where the stone could be brought and dressed. A vessel was moored about half a mile from the rock as a base for the seamen and the masons, who would work one week on the rock alternating with one on shore. The rock was soon a busy building site with lifting gear in place, and workers cutting away the rock to create a base for the stonework. The scale of the operation was immense: for the base layer 1,493 individual pieces of stone – a total weight of around 1,000 tons – had to be cut to shape and shipped out to the rock. The work progressed steadily under Jessop's supervision, and by 9 October 1759 all was complete; a week later the light shone out over the sea for the first time. It was a triumph – and the success of the design meant that it was to be a model for many other lighthouses that were to be built over the coming years. It remained in use right through to the 1870s, when a new, higher lighthouse was constructed and Smeaton's dismantled. The upper section was re-erected on Plymouth Hoe, where it remains and is simply known as Smeaton's Tower. There was one other important outcome from the work on the lighthouse: Smeaton agreed to take on Josiah Jessop's son, William, as an apprentice. We shall be meeting him again shortly.

Smeaton did not spend all his time at Plymouth, and was able to take on other work, including reporting on harbour schemes and doing more scientific experiments. The Eddystone project brought him a secured income for some years, enough to be able to get married, but not to his first love. His bride was Anne Jenkinson, the daughter of a merchant tailor and freeman of the City of York. There is no record of when they first met or became engaged, but they were married in London in June 1756. Among the tasks he took on was a survey of the River Calder.

The river has its source in Lancashire and flows down to join the River Aire near Castleford. In 1699 an Act of Parliament authorised improvement of the Aire to Leeds and also of the Calder from Castleford to

Wakefield. A number of locks were built, 60ft long and 15ft wide, and the Aire & Calder Navigation was opened throughout in 1704. Smeaton was now investigating if navigation on the Calder could be extended to somewhere near Halifax. He was not the first to look at the problem: there had been two previous surveys in 1740 and 1741, but nothing had been done. It was a daunting task, for the river had a fall of 178ft in just 24 miles. To put that in context, it represents a fall of 89in per mile, and the steepest fall previously dealt with was on the Kennet at 65in per mile. To compound the problem, there was always the chance of flooding when heavy rainfall swept over the Lancashire hills. The river passed through an area famous for woollen textile manufacture, so there were fulling mills along the route with accompanying weirs – and they would have to be bypassed.

Smeaton's plan was to construct twenty-six locks with 5¾ miles of artificial cutting and extend the navigation a little way up the River Hebble to Brooksmouth near Halifax – this was later extended up the Hebble to Sowerby Bridge. It would be called the Calder & Hebble Navigation. He planned to build a reservoir at Brooksmouth, and carry out dredging to remove shoals on the Calder. The Act was passed in June 1758 and Smeaton was appointed as chief engineer, with Joseph Nickalls as resident engineer. He was plagued throughout by differences of opinion among the proprietors and, although Smeaton had complete confidence in Nickalls, he was dismissed in 1761 and replaced by two men, one to be in charge of carpentry, the other masonry. River navigations involved making artificial cuttings, with locks in them, to bypass existing weirs, built by mill owners, or by constructing both locks and weirs to overcome differences in level. Smeaton had also recommended installing flood gates at various points and suggested some banks needed strengthening. By July 1764 the section from Wakefield to Brighouse was completed with a total of seventeen locks in 18 miles, and Smeaton could reasonably have expected to be asked to complete the work. Instead, the job was offered to James Brindley, who accepted. There was no reason given for the change, but one can hazard a guess: thanks to the publicity surrounding the opening of the Bridgewater Canal, Brindley was the man of the moment. Everyone, it seemed, thought he was the best you could get to build your canal, and the Calder & Hebble proprietors were no different.

Everything was proceeding well, until the night of 7 October, when storms sent a great flood crashing down the river, causing immense damage and bringing all traffic on the navigation to a halt. Now the proprietors turned back to Smeaton for advice and he agreed to return to survey the damage and suggest what needed to be done. He could not resist pointing out that, had his original suggestions for preventing flood damage been carried through, the disaster might have been avoided, or at the very least the damage could have been considerably less. Unfortunately, there was no money available for repairs, and the company had to go to Parliament to get a new Act to authorise issuing shares. This was granted in 1769 and the repair work carried out. The whole line reopened to Sowerby Bridge, but with no further involvement of Smeaton. He, however, now had other canal and river schemes to consider.

Smeaton was never expected to be present throughout the construction of the Calder & Hebble – he was the consulting engineer; the resident engineer had the job of overseeing everyday activity. This left him plenty of time to act as consultant to a variety of other schemes. John Grundy had surveyed for a canal from Louth in Lincolnshire to a point near Tetney Haven on the Humber in 1756, but nothing had been done. Smeaton was brought into advise on the scheme. It was comparatively simple to build a canal through the flat land starting with a sea lock at the basin, after which there was 9¾ miles on the level, but there was a rise of just over 54ft in the remaining 4¼ miles. The Act was approved in 1763, but with no further involvement by Smeaton.

The Fossdyke has the distinction of being Britain's oldest canal, having originally been constructed by the Romans, but much altered over the centuries. Now it had fallen into disrepair and Smeaton was among those who were consulted with Grundy on what should be done. Various improvements were suggested, including building a sea lock at Boston, dredging the navigation and building three new locks. Other reports included a scheme for making the Upper Wear navigable, which was approved, but nothing was ever done; and a report on the River Chelmer, which was eventually worked on, but only in the 1790s. He was also approached by the Wey Navigation Company in 1760, but he had to turn them down. He had 'a large affair in the North' to occupy him. This was the proposed Forth & Clyde Canal.

The story of this canal is complex, simply because of the variety of different schemes that were put forward. Smeaton's earlier suggestion for deepening the Clyde had been partially carried out thanks to the Clyde

Act of 1759, and now it was proposed to link the river to the Firth of Forth. There had been a survey as early as 1726, but no further work had been undertaken: then in 1762 there was another by Robert Mackell and James Murray. The following year, Smeaton was invited to make his own recommendations on the alternative routes.

Although by this date, the Bridgewater Canal had been built, the first in Britain that was independent of any natural waterway, the Forth & Clyde was a very different proposition. The Bridgewater had been supplied with water from the mine workings at Worsley that, incidentally, was a subject in which Smeaton had been involved briefly. The question was: would the mines supply enough water or would water have to be supplied from nearby rivers and streams, affecting local mills? He devised a simple instrument, with Brindley, to measure water flow and was able to report that the mine supply was adequate. The Bridgewater ran on the level, until an extension was built to the Mersey, where locks were constructed. The situation on the Forth & Clyde was very different, as it would be what could be called a summit or watershed canal. It would have to rise up at either end from the two rivers to a higher summit. Every boat using the canal would drain away two locks' worth of water. A boat leaving the summit would fill the lock, then empty it: the water would pass down to the next lock, but would eventually be lost in the river, either to the Clyde for a boat travelling from east to west, or into the Forth for boats travelling the other way. This would have to be supplied by some means or other to prevent the summit level running dry.

No canal of this type had yet been built in Britain, although a major summit level canal, the Canal du Midi, already existed in France, which actually had many other engineering features yet to be seen in Britain, including the world's first canal tunnel at Malpas. But although Smeaton had looked at canal work in the Netherlands, he had not seen the French canal. He had to come up with his own scheme. Two possible routes were considered. The first would involve a canal that would leave the Forth near Stirling, then cut across country, roughly along the line of the present A811, to Loch Lomond, and then use the River Leven to join the Clyde at Dumbarton. The alternative that Smeaton favoured was a far more direct line that left the Forth near the village of Carron, which was also home to a great iron works that would certainly be providing trade for the canal, and would join the Clyde near Yoker, downriver from the centre of Glasgow.

Smeaton's plan was very much designed with water supply in mind. There were to be seventy-seven locks in the first version, able to take the vessels then in use on the Clyde, which were 56ft by 17ft. In order to conserve water, he stipulated that the fall at any one lock should not exceed 4ft, and the summit level should be cut deeper than the rest of the canal to act as a reservoir. He did, however, put in a note to say that if water supply proved better than expected, the number of locks could be reduced. Major engineering works included three aqueducts and a short tunnel. Smeaton drafted a full and detailed report, but it was bedevilled by arguments among the different interested parties and was never acted on. A new survey was carried out, involving Mackell again, this time accompanied by another of our engineers, James Watt. This was for a much less ambitious canal a mere 24ft wide to take small craft. In 1767, this proposal went to Parliament, but was opposed by another group that favoured the much larger canal first proposed by Smeaton. Eventually, the Bill for the small canal was defeated, and Smeaton was once again required to produce yet another plan for a canal to take vessels 60ft by 20ft.

It was clear that water supply would now be more of a problem, so he proposed building a substantial reservoir at Townhead and widening the canal at Dullatur Bog. With supplies secured, the canal only needed forty-two locks, each 70ft by 20½ft. The Act was approved in 1768 and everything seemed settled. Smeaton was appointed chief engineer and work could get under way. Then, incredibly, yet another proposal was put forward, promoted by the Carron iron company, with Brindley proposing a narrowboat canal. Smeaton was unimpressed and scathing in his comments:

As no difficulty is too great for Mr Brindley, I should be glad to see how he would stow a fire engine [steam engine] cylinder cast at Carron, of 6½ft diameter in one of his seven foot boats, so as to prevent it breaking the back of the boat, or oversetting.

The narrow canal option was dropped, but there were to be major changes to the route, which would eventually run from a junction with the River Carron. The river was straightened to its junction with the Forth to a basin beside the Clyde at Bowling. There was also to be a short branch to Port Dundas, a mile north of Glasgow city centre. The Townhead reservoir, now usually known as Barton Loch, was a massive undertaking and covered

54 acres, contained behind a 25ft-high dam. Work began at Grangemouth on the Forth under Smeaton's direction, with Mackell as resident engineer. By 1773 the canal was opened from the Forth to the summit at Kirkintilloch. At this point, Smeaton resigned, happy to leave the job of completing the work to Mackell. Mackell died in 1779, and the company asked Smeaton to come back again to take over, but he declined, suggesting that it would be better to bring in a new man. So, the task of completing the Forth & Clyde went to Robert Whitworth, and he had the honour of designing the most imposing feature of the canal, the Kelvin aqueduct.

Smeaton was justifiably proud of what had been achieved. A Russian engineering officer was sent to Britain to see what was being done with canal construction, and the Russian Consul wrote to Smeaton for suggestions: his answer was to go and see the Scottish canal – 'they will get more real knowledge by considering the Construction of this Canal than all the rest put together.' He was soon being consulted on other navigation and canal schemes.

3.3 Port Dundas on the Forth & Clyde Canal in the nineteenth century. (From Alfred Barnard, *The Whisky Distilleries of the United Kingdom*, 1887)

The River Lea has a long history of river improvement, starting with an Act of 1425, and in the 1570s one of the very first modern locks with mitred gates was installed at Ware. By the middle of the eighteenth century, however, the river was far from being an up-to-date waterway and still had many of the old-fashioned flash locks in place. The commissioners called in Smeaton to come up with proposals for a modern waterway. He surveyed the river with Thomas Yeoman to help, and came up with a proposal that included shortening the river navigation by artificial cuts and twenty-one locks, and he also suggested an extra cut from the river down to the docks at Limehouse on the Thames. An Act was obtained in 1767 and work went ahead with Yeoman in charge. He was to return to Yorkshire for a number of schemes, including work on the Upper Ouse and the Ure, with a short canal to Ripon that was completed in 1773. The same year, he went to look at the works on the Grand Union Canal of Ireland and wrote reports on what needed to be done. We shall be looking at the results of that visit in the next chapter.

We have moved some way ahead in the Smeaton story, because although his work on navigations and canals was important, other things were going on in his life at the same time. The family had suffered the loss of their first daughter, but sometime in the early 1760s they moved north to the old family home at Austhorpe Lodge, following the death of Smeaton's parents. Two other children were born there: Mary in 1761 and Hannah in 1765. With a growing family, the loss of the Calder & Hebble appointment was a financial blow, but Smeaton was fortunate in being recommended for the vacant post of receivership of the Greenwich Hospital Derwentwater Estates. The other official receiver looked after the day-to-day running of the extensive estates that stretched across much of the north of England, so this was something of a sinecure for Smeaton. He only had to put in an appearance for a few weeks once a year, for which he received a handsome salary of between £400 and £500 a year. He was soon, however, to be kept busy on a whole range of projects, and one type was constant throughout the rest of his career: designing mills of many different kinds.

Over the years, Smeaton designed around sixty mills of various kinds, mostly watermills, though the list includes four windmills and some oddities. For Kew Gardens, in 1761 he produced a device for raising water that used two horses to turn an Archimedes screw and, much later, in

1789, he supplied a two-man tread mill to work a crane at the London Custom House. The watermills also showed great variety and some very new features. But before any work could be carried out, or design work begun, Smeaton's first task was to ensure that there was enough water to turn the wheel and then work out, given the fall, what type of wheel to design – undershot, breast shot or overshot. Traditionally, millwork had been constructed entirely of timber. Smeaton was among the first, if not the first, to start making extensive use of cast iron. He used metal for the cogs, for which he had prepared wooden patterns from which they could be cast. He used iron axles for water wheels on a few mills, and discovered that iron buckets for overshot wheels were more efficient than wooden. The use of the heavier metal on water wheels for the rings that held the paddles was also useful as they acted as a flywheel that would ensure smooth motion.

Smeaton's interest in using iron came from his involvement in providing machinery for a number of iron works, including one of the country's most famous at Carron. They are best known for the short-barrelled cannon they built for the Royal Navy, known as carronades. At the time there were four furnaces to be brought into use and the temperature for smelting had to be raised by blowing air in with blowing machines. In a blacksmith's forge, air could simply be blown by foot-operated bellows, but on the larger scale of the furnaces, the blowing machines had to be worked by water wheels. Smeaton was faced by a complex problem. The four furnaces were not all the same size, so he eventually finished up building two water wheels, one overshot and one high breast shot. He would later add another wheel for operating a boring machine.

One of Smeaton's most ambitious projects was for the pumps powered by a water wheel set in one of the arches of the old London Bridge, which was noted for the fast flow of the river through the narrow openings. He designed an undershot wheel, 32ft in diameter and 15½ft wide, which drove three pumps to either side of the wheel. Each pump rod was suspended from one end of an overhead rocking beam. The other end of the beam carried a connecting rod attached to a crank, turned via gearing from the water wheel. The arrangement can be seen from the illustration, taken from Rees's *Cyclopaedia*.

Smeaton was one of several engineers who looked for ways to make improvements in the Newcomen engine. At this time, the standard

3.4 Diagram
showing the
waterwheel and
pumping mechanism
for the Thames
waterworks.

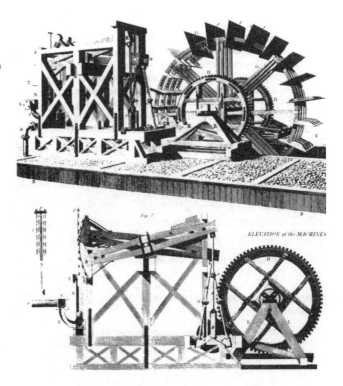

measurement of efficiency was 'duty'. This was the number of pounds of water that could be raised a foot by the consumption of a bushel of coal. The duty of the early Newcomen engines was roughly 4.5 million. The first engine to Smeaton's design that was actually built was to pump water for the New River Company of London. He identified one problem as being that the beam was not in continuous movement as it stopped when the pumps were lowered and there was a pause until the steam had been evaporated, and then another pause when the rods had been raised while pressure was equalised and gravity could once again get to work. Overcoming inertia took up a lot of the available energy. Smeaton decided that having larger pumps with a longer stroke and a slower movement could improve efficiency. Eventually he got the engine working at 7 million duty. There were, however, still improvements to be made.

Typically, Smeaton set about the matter scientifically. He first gathered information on engines then at work to see which were the most efficient and try to identify why. In general, the bigger engines, where the cylinder diameter could be as large as 74in, performed better than the smaller ones, under 40in. He then developed a small test engine with a 10in cylinder at Austhorpe. He set it to work in 1770 and measured its efficiency. Then he would alter one element and leave everything else unchanged to see if it made a difference. Altogether he carried out 130 tests until he was convinced that he had the best combination of all the different factors.

Armed with the results of his experiments, he was able to design a pumping engine for the Long Benton colliery. There was an engine already on site that was rated at 4½ million duty. Smeaton's engine had a 52in cylinder, the same as its predecessor, but it was rated at just over 9 million duty. Not surprisingly, Smeaton was soon being asked to devise engines for other mines. One of his more ingenious designs was for a winding mechanism for lifting the coal from the pit at Long Denton. The mechanism itself was powered by a water wheel, but in order to keep it regularly supplied with water a small steam engine was installed to pump the water back after

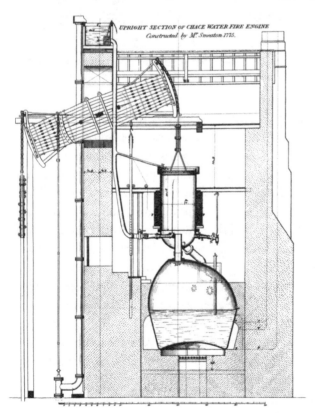

UPRIGHT SECTION OF CHACE WATER FIRE ENGINE
Constructed by Mr Smeaton 1775.

it left the wheel. It was built in 1777 and over the next decade two more similar engines were built at the colliery.

3.5 Smeaton's improved Newcomen engine built for the Chacewater mine. (D.B. Barton)

Smeaton's engines were among the most powerful of their time, but in spite of all his scientific experiments he had failed to spot the fundamental flaw in all Newcomen-type engines. That is a subject we shall return to later. He could, however, claim to have brought the Newcomen engine to its maximum efficiency. As well as his work as a mechanical engineer and millwright, Smeaton also had many important civil engineering projects in a busy working life, which can be roughly divided between bridge building, harbour construction and, to a lesser extent, fen drainage.

Smeaton approached the business of bridge construction in the same methodical way that he had when considering steam engines. He thought carefully about such matters as the ratio between the thickness of an arch and the distance it had to span. He needed to work out the lateral pressure that the arch would exert, which, as he wrote in a pamphlet, had never been accurately measured before. Once he had the solution to that problem, he wrote that he – he referred to himself in the third person – 'has given such a size and form to his middle piers as will resist several hundred tons more lateral pressure than the middle arch will exert thereon'. His first project was for a bridge across the Thames at Blackfriars, for which he produced designs and estimates, but eventually the contract went to Robert Mylne. He also designed a bridge for Glasgow that was never built. His first major work to be completed was for a bridge across the Tweed at Coldstream. This is an elegant bridge, built between 1763 and 1767. It crosses the river on five semicircular stone arches, each just under 61ft span. One of the features of the bridge is the comparative narrowness of the arches, compared with those of other masonry bridges of the time, being just 2½ft wide. Single-walled coffer dams were used to provide a dry area in which to construct the foundations. However, Smeaton had a problem with the loose gravel of the riverbed through which the piles were driven and water kept seeping through.

His next major bridge at Perth, designed in 1764, presented a very different problem when it came to the foundations: Smeaton found the gravel bed was so densely packed he could only drive the piles in to a depth of around 2ft. To keep the workings dry, he packed the bottom of the dam with earth and gravel. The design is very similar to that of Coldstream, but with 75ft span and arches consequently thicker at 3ft. His third Scottish bridge at Banff was of very similar design to the other two, but smaller. His other major bridge was built at Hexham, and he followed exactly the

3.6 Perth bridge. (From *The Imperial Gazetteer of Scotland*, 1868)

same principles as he had with his previous bridges. Work was completed in 1777. Then in March 1782, flood water swept down the Tyne with far greater force than Smeaton had ever envisaged. The bridge was demolished. It was a bitter blow to the engineer, who wrote: 'All our Honours are now in the Dust!' It was his first and last major failure. It was also to be the last major bridge work in which he was involved. His one major contribution to bridge construction was that his bridges were generally built using far less material than those of his contemporaries. He never followed other engineers of the day by using iron in the construction.

At the same time as he was working on bridges, he was also engaged in a number of harbour projects. His first was for a new pier at St Ives. The foundation would be based on *pierres perdues*, mounds of loosely packed stones. Above that was the pier itself, 360ft long, and rising 36ft above the foundations, with a 9ft parapet. It has now been extended, but is still known as Smeaton's pier. His next job was for a pier at Eyemouth. A far more important project was the construction of a harbour at Portpatrick on the west coast of Scotland. A quiet place now, it was then a busy port providing a short sea crossing to Ireland, but was only provided with a basic landing stage and no real protection from bad weather. His plan of 1770 called for a 270ft-long pier extending out from the shore, and a second pier built near the end of the main pier and at right angles to it. The scheme was later extended.

When he came to look at proposed improvements to the harbour at Aberdeen, he was faced with a quite different problem. The entrance to the river was partially obscured by a sandbar. Fortunately, this was a phenomenon he had already studied, and he was ready with an answer. He proposed to build a 1,400ft-long pier along the north side of the river, which would prevent sand and gravel being blown from the land into the river. The scheme was successful up to a point as it now became a simple matter to keep the channel dredged. However, once the bar was lowered, there was a new problem, with easterly winds carrying waves further up the river. To counteract this, Smeaton designed a short pier at an angle to the main north pier. This diverted the waves to the opposite shore, where they broke on a sandy beach. At the same time, he recommended improving the sea defences along the southern bank. By 1790 the whole of the works had been completed. His other works in Scotland included piers at Peterhead and Cromarty. There was, however, one other major harbour project that occupied a great deal of his time, and it began when he was asked to go to Ramsgate to advise on how essential improvements could be made.

The problem was that although protecting piers had already been built, silt had built up to an alarming extent, so that the harbour was becoming totally clogged up. Smeaton looked at two possible solutions: the first was to dredge the existing harbour and keep it dredged, but with an estimated 400 tons of sand and gravel being washed in every week, he decided that was impractical. Instead, he suggested building an inner basin, which could be emptied through sluices that would act to wash away the silt. The plan was later modified by the construction of a wall with a pair of gates, running right across the harbour. Work began on this proposal in 1776. Altogether six sluices were built and proved effective. He continued with other projects at Ramsgate, including a new pier and a dry dock. He was visiting Ramsgate to supervise the work right through to the spring of 1792. What is notable about Smeaton's harbour works is that each seemed to present a different problem, to which he always found a good and economic solution.

Smeaton's very first report on drainage was for the Lochar Moss near Dumfries, and it was completed as early as 1754. It was an ambitious scheme, but was never actually carried out. The next scheme on which he was consulted was for the drainage of Potteric Carr, an area of around 1,700 hectares just south of Doncaster, now crossed by the M18. 'Carr' is a name derived from the old Norse word meaning fen. He put forward two proposals. The first was a drain that would pass through the ridge carrying the road from Doncaster to Balby (now the A630) via a short tunnel, and would eventually drain into the Don near Friars Gate Bridge, where the Don Navigation and the river surround a small island at Doncaster. The alternative scheme involved a drain that would lead eastward to join the little river Torne near Rossington. There seems to have been a problem with the route to the Don as it appears there was a miscalculation in the difference in levels between the Carr and the river. In any case, a second survey was carried out, but no details have survived. Smeaton then proceeded to do a more detailed survey of the second option that now included straightening the Torne at Rossington to ensure that water from the drain could flow away easily. The fall from the Carr to the river was a modest 5in to the mile, so Smeaton designed the drain to be broad enough to ensure a good flow, with a width at the bottom of 18ft. The scheme was adopted and completed in 1777.

The Adlingfleet drainage scheme was designed to prevent the regular flooding of some 200 hectares of land to the south of the Ouse and to the west of the Trent where the two rivers meet. Existing drains had proved inadequate, being both crooked and narrow. One solution would have been simply to straighten and widen them and Smeaton gave estimates. He also put forward a proposal for a new main drain to reach the Trent at Trent Falls, close to the confluence with the Ouse. Although it was more expensive, the latter course was adopted and the work was completed in 1772.

The next major project took him some 6 miles south to Hatfield Chase. The area had already been tackled at the beginning of the seventeenth century as part of the huge fen drainage system undertaken by Sir Cornelius Vermuyden, but much of it was now inadequate. One of the most important drains was the Snow Sewer that took water off the low lands south of the Isle of Axholme to an outfall at the Trent 6 miles away. Smeaton found that when the sluice doors were shut by the tide, the water in the drain simply overflowed. He proposed a new sluice to overcome the problem. Another problem was the regular flooding of the River Torne.

The course of the river had been straightened by Vermuyden, but still presented problems. Various schemes were proposed over the years, but it was left to Smeaton to come up with a comprehensive plan for the area, involving new sluices and improving the old. Fen drainage may have been a small part of Smeaton's professional life, but the work he undertook was of real value for the region.

Throughout his career, Smeaton never neglected his scientific interests, whether conducting his own experiments or visiting the Royal Society to hear about the latest theories and developments. We mentioned earlier about his experiments on water wheels, which were of fundamental importance in the many mill projects in which he was involved. Although, he only designed four windmills, one of which we know virtually nothing about apart from its location at Barking, he did carry out a whole set of experiments to decide on which sails were the most effective and other factors such as the best angles for the sails to be set. He built a model with which he could test for the different variables. This consisted of a sail set at one end of a long rod, counterbalanced by a weight at the other, and mounted on a spindle, wound with twine. By pulling the twine, the spindle rotated and the arm moved round, causing the sail to move. This was attached to a pendulum and each rotation of the sail corresponded to two beats of the pendulum. He was able to vary the 'wind speed' from an equivalent of 4ft 4in to 8ft 9in per second. As with the water wheel experiments, he showed that previous authorities had got it wrong: for example, one authority, Antoine Parent, had stated that the angle that the sails should be set to the wind should be 35 degrees, but Smeaton found that an angle of less than half that produced maximum efficiency. In another experiment, he found that increasing sail size improved efficiency up to a certain point, beyond which it fell away. In general, his work on windmills was not as definitive his study of watermills. In the latter case he decisively proved the greater efficiency of the overshot wheel: nothing quite that basic emerged from the other study.

His study of the laws of motion and mechanical power might seem more abstract. The basic laws of motion first set out by Sir Isaac Newton are so generally accepted that it seems difficult to imagine they were ever seriously challenged. But they were by another great scientist, Gottfried Leibniz. Once again, Smeaton set up a mechanism to investigate mechanical power. His results showed conclusively that Newton had the right answers. In speaking of power, he worked out that it could be expressed in precise terms, involving the variables of mass, velocity and time. This is basic physics, but for Smeaton it could also be expressed in simple practical terms. He used the analogy of a watermill, where a certain amount of water is available for grinding: it will only grind a specific amount, which will be independent of how fast the water flows; a quick flow and the job will be done sooner than with a slow flow, but the amount will always be the same. There had been disagreement between Leibniz and Newton on a definition of momentum – which in Newton's version was proportional to velocity, and in the latter to the square of the velocity. In terms of pure physics, Smeaton demonstrated experimentally that Newton's definition of momentum was correct, and that work done was proportional to velocity not momentum. In another set of experiments, he looked at the energy lost in impacts between non-elastic bodies.

He continued with various other experiments of practical value, including work on the loss due to friction in liquids flowing through pipes of different dimensions. He set out to measure how much work could be done by horses or men in jobs such as pumping up water or raising weights. He came to an eventual conclusion that a horse in an eight-hour day could work at the rate of 22,000ft lb per minute – foot pounds being a measure of the product of the weight multiplied by the distance moved. The idea of using the work done by a horse as a standard measurement of power was only developed later. He also experimented with hydraulic limes, essential components for example in the construction of the Eddystone Lighthouse. In all his scientific work, he showed a sound approach to experimentation as the basis of knowledge, but seldom lost sight of the practical implications of what he was achieving.

Smeaton was among a select number who established the role of the civil engineer as a profession rather than a trade. This was formalised in 1771 with the formation of the Society of Civil Engineers, of which Smeaton was a member. It was a dining club rather than a professional body, but it was an opportunity for members to meet and exchange ideas. In later years, members became known simply as the Smeatonians, emphasising his central role in the development of the profession.

Smeaton's personal life was hit by tragedy. His daughter, Hannah, died in 1776, just 11 years old. Another daughter, Ann, however, lived on and was married in 1780 to John Brooke, a linseed oil manufacturer.

The following year Smeaton fitted out a windmill for grinding the linseed. They lived at Sykefield, close to Austhorpe Lodge. His other surviving daughter, Mary, married Jeremiah Dixon in 1781, initially living in Yorkshire – he was to be mayor of Leeds – but later settling in the Lake District. In 1783, Smeaton's wife became seriously ill and, in spite of a spell in Bath, where it was hoped she might be cured, she died on her 59th birthday in January the following year. John and Ann Brooke now moved in to look after the Austhorpe household. Smeaton continued working on his many projects until 1791, when he made a general announcement that he was retiring from active engineering work and was to spend his time writing up reports on his various activities. However, he suffered a stroke at Austhorpe in September 1792 and died six weeks later on 28 October.

Smeaton's career was extraordinary in covering such a wide range of disciplines, and in particular for his use of scientific experiments to solve practical questions. The one outstanding achievement, however, has to be the Eddystone Lighthouse. Where other activities such as millwork and bridge construction represented no more than variations on existing practices, his lighthouse was completely novel and set the pattern for future buildings around the coast of Britain. By publishing reports and papers, he made his own work widely known to other engineers and he passed on much of his expertise to his young apprentice, William Jessop, who is our next subject.

4

WILLIAM JESSOP

William Jessop was born in 1745 and he was just 14 years old when he was taken on by Smeaton as an apprentice. As was usual, the family paid a premium to Smeaton to cover the costs of tuition and, when Smeaton left Plymouth, young William went with him to live at Austhorpe Lodge. When William's father, Josias, died in 1761, his mother struggled to keep up the payments and, in 1764, Robert Weston, the major shareholder in the Eddystone Lighthouse project, who must have appreciated the Jessop contribution to the completion of that great work, took over the payments until the apprenticeship ended in 1767. At that point, Smeaton took Jessop on as an assistant.

Jessop was soon being sent on his own to provide estimates for work, which Smeaton was too busy to attend to in person. The first job for which he received personal credit came when Smeaton was working on schemes to improve navigation on the Aire & Calder. One proposal was for a long cut with two locks between Chapel Haddlesey and Newland. As Smeaton was looking at various other parts of the system, he got Jessop to provide the plans and estimates for this section. He was able to present it complete with a new title – Mr Jessop, engineer. He was now 27 years old. The section was, however, never built and a different route chosen. Before that, however, far more important work came his way.

In the previous chapter, there was a mention of Smeaton going to Ireland to look at the works on the Grand Union Canal that had been languishing unfinished for years. When accepting the offer, he had told the Irish proprietors he would not be able to do more than conduct an initial survey of the route, but that he would bring with him 'a young Gentleman who has just begun Business for himself'. This was, of course, William Jessop. He carried out a study of levels for the new section in 1773 and at the same time was asked to go and report on the Tyrone Canal, which had small tub boats and instead of locks used inclined planes to overcome the differences in levels – the first of its kind in the British Isles. The system used rollers, and the weight of the loaded boats was used to raise the empty ones back to the canals' upper levels. But the rollers kept sticking and Smeaton suggested substituting railed tracks for the rollers. Jessop went again to report on the scheme, and eventually it was modified: instead of sliding down the rails, the tub boats were floated onto wheeled cradles on the tracks, and power was supplied by a horse gin. Whether Jessop was involved in suggesting this improvement or not we shall never know, but we do know that the idea remained with him and was later put to use. There was one other major event in this period of his life: he got married to Sarah Sawyer in 1771. Their first child – a son, John – was born two years later. Altogether they would have six sons and one daughter.

The first flood of activity in canal construction following the success of the Bridgewater Canal slowed to a trickle, largely because money was short during the American war. In a way, this was fortunate for Jessop, and there was still work to be had, especially in improving the river navigations around the country. The experience he gained there and the reputation he gained for reliability would make him the ideal man to take on major projects once prosperity returned. He reported on work needed on the Soar Navigation and he had a major role in improving the River Trent. An earlier report had proposed the traditional remedy of avoiding shallows by means of cuts and locks, but that was not accepted, largely on the grounds of cost. In 1784 Jessop carried out a survey and found a total of sixty-seven shoals that needed to be dealt with, and he put forward a number of suggestions on how best to tackle the problem. The one he favoured was for making a navigable channel, 30ft wide and 18in deep, with a towpath for horses. This was agreed, but was later amended and the depth increased to 24in. The work was approved and completed in just six months in 1784. He was to be employed regularly as an engineer to the Trent Navigation, and the family moved into a house in Newark. Other work followed, on the Shropshire tub boat canal and the Soar Navigation again. His next major assignment was to take him to the Thames.

The Thames had scarcely seen any improvements since John Taylor, the waterman poet, wrote an account of a journey down the whole river in 1632, describing in verse the atrocious conditions he found. Shallows caused many problems: here is a typical description:

> At Clifton, there are rocks, and sands, and flats,
> Which made us wade and wet like drowned rats.

In place of the familiar locks we know today, there were mostly flash locks. In these, differences in levels were overcome by building a weir with moveable paddles. Water would build up behind the weir and, when boats needed to pass, the paddles were removed. Boats going downstream went down on the flash of water, like shooting the rapids, while those going upstream were winched against the flow. It could be decidedly dangerous:

> Then Sutton locks are great impediments,
> The waters fall with such great violence.

Indeed, they did represent a genuine threat: at Marlow a passenger boat capsized in a flash lock and all on board were drowned. There were three pound locks, which Taylor referred to as 'turnpikes'. This was much the situation when Jessop arrived at the river a century and a half later. There were still only three pound locks on the whole length from Inglesham to Mapledurham. He put in hand several improvements, including a new lock built out of timber at Whitchurch, 2 miles above Mapledurham. He suggested replacing the flash lock on the upper Thames above Oxford at Nuneham Courtenay, but that was not agreed and the existing lock remained in use right up to the twentieth century. Another obstacle was the narrow bridge at Radcot, and Jessop's plan to bypass it by an artificial cutting and lock and a new bridge was approved.

He surveyed the whole of the allegedly navigable river from Inglesham to Oxford and was singularly unimpressed, finding it very crooked and choked with weeds. The crookedness makes it seem charming to modern pleasure boaters – Eric de Maré aptly described it as being 'like a stream in a missal' – but hardly suitable for heavily laden barges. Jessop's proposal, however, was not to cut across the bends, but to dredge a passage through the shallows and use the excavated material to build up a towpath. He also designed a number of new locks below Oxford, this time built of stone rather than timber. He worked on a number of other projects at this time, including the Ipswich and Stowmarket Navigation and the Sussex Ouse. They were not major undertakings, but everything was about to change. The nation was recovering from the depression caused by the American wars and canal construction was soon to become so intense that the period came to be known as the years of canal mania.

The first major canal on which Jessop had a major role to play was the Cromford. Originally this was planned to run from the Erewash at Langley Mill, Derbyshire, and then to connect to the Derwent, which would be made navigable. Later it was decided not to use the river, but to extend the canal to Cromford, where Richard Arkwright had established his famous cotton mill. This would involve a 3,063-yard tunnel at Butterley. The canal from the Erewash to the tunnel would be broad to take the river barges, but the tunnel itself would be narrow, just 9ft wide. There were to be two large aqueducts, one over the Amber and the other across the Derwent. The Act was passed in 1789 and Jessop was appointed chief engineer, with Benjamin Outram as the resident engineer.

At first things went well, but the main contractors left the works in 1791, having been offered a more lucrative job on the Glamorganshire Canal. They had been overpaid by more than £1,000. Worse was to follow. The Amber aqueduct was an imposing structure of three arches, rising 50ft above the river. In January 1792, the structure failed. Jessop at once accepted full responsibility and offered to pay £650 to cover repairs and give up his salary until the canal was completed. The £650 was accepted, but the company continued to pay his salary of £300 pa. Then in 1793 a crack appeared in the slightly smaller Derwent aqueduct, which Jessop put down to having used poor-quality lime and attempting

4.1 The Derwent aqueduct and chimney of the Leawood pumping station on the Cromford Canal. (Author)

to economise in his use of material. The aqueduct was repaired and still stands today. The tunnel was the other major undertaking, with a total of thirty-three shafts being sunk and a steam engine installed to pump out water. Work on that went altogether more smoothly.

One thing that was clear from the work on the tunnel was that the area was rich in iron ore and coal. Francis Beresford, a local solicitor, bought the Ripley estate near the canal in 1790 with Outram as a partner. A limestone quarry was opened at Crich, now home to the famous tram museum, and that was connected to a wharf and lime kilns by the canal. Mining began at once and, shortly afterwards, a blast furnace was built at Butterley. Jessop was later taken into the partnership. The Butterley iron works were a huge success and were to play an important part in several of Jessop's later projects.

Jessop continued to do work for the Trent Navigation, and one problem he was asked to consider was how to deal with the difficult passage through Trent Bridge in Nottingham. The issue was complicated by the fact that a number of other schemes involving junctions with the Trent were under consideration at the same time, including the Derby Canal and the Nottingham Canal. The Nottingham Canal was authorised in 1792 to run from the river to join the Cromford Canal. This provided an opportunity to build a cut from the river at Beeston to the canal and then to rejoin the river below the bridge. Jessop continued making several more improvements to the Trent.

His next project took him to neighbouring Leicestershire and the River Soar. There were two separate elements. The first was an improvement of the river itself near Loughborough, by the construction of several cuts to bypass difficult sections, of which the most significant took the canalised section from a point near Barrow upon Soar for 3 miles to rejoin the river beyond the Loughborough wharf. It was then proposed to have a branch that would head west towards Swannington. The first plan was for a conventional canal with locks, but it was later decided that, as it approached the hillier region at the western end, the canal should come to a halt and the route be continued by a tramway, a railed track, with horses for haulage. As a result, the canal itself could be kept on the level, but only by taking a very convoluted line. Known as the Charnwood Forest Canal, it was not a great success. Indeed, looking at the OS 1:50,000 map of the area, not a trace can be seen. One feature does, however, remain. To supply the canal with water, the 33-hectare Blackbook reservoir was constructed at Jessop's suggestion, though the actual building was left to Christopher Staveley. It was completed in 1797, but just two years later, in February 1799, the earth bank gave way and in just eleven minutes the whole of the contents had spilled, flooding the surrounding area as far as Loughborough. The canal never recovered from the catastrophe, but the reservoir remains, now a nature reserve.

Jessop advised on several projects in the area, including the Melton Mowbray Canal. He was not paid for his initial survey of 1785 and had to sue to get the money. The Act was passed in 1791. He did not have the same problem with the Nottingham Canal, which was to form a connection between the Trent at Nottingham and the Cromford Canal. Although Jessop was nominally in charge, he had very little to do other than approve the plans and the survey. He was certainly kept busy advising on a whole variety of schemes in the region: the Derby Canal, the Ashby Canal, the Leicestershire & Northamptonshire Union Canal, the Grantham Canal, the Witham Navigation, the Sleaford Navigation and the Horncastle Canal. It is hardly surprising that he felt unable to offer his full services as an engineer to such a collection of projects.

His working life in the late 1780s and early '90s was not limited to canals and rivers. Like his mentor, Smeaton, he took on a number of land drainage projects in East Anglia, and it was Smeaton who recommended him for the first project in 1784. It was a modest task, building a new sluice for the fens south of King's Lynn, but when that was carried out satisfactorily, Jessop was invited to investigate a far more important project. The commissioners responsible for land drainage wanted to know if it was possible to use steam engines for land drainage in the area. Jessop had a long consultation with Boulton & Watt, and between them they decided that for the main area of 5,000 acres a 40hp engine would be needed to lift the water by 8ft. Nothing came of the proposal at the time. One other project required a report from Jessop in the region for a cutting across an extravagant bend on the River Ouse just below King's Lynn: the Eau Brink Cut. Jessop approved the plan, but although an Act was approved in 1795, nothing happened until 1818.

At the same time he was looking at the Holderness area in East Yorkshire, for which he proposed new drains and alterations to existing ones. His report of 1786 was accepted and he was later asked to report on ways of improving the flow in the River Hull. He was to be involved in other drainage schemes later, but by the end of the 1780s he had a major canal project to look after. He was to be busy again in Ireland.

The first visit to the Grand Canal with Smeaton was described in the previous chapter. He was back again in 1789 to make a report on progress, and by 1790 he was spending two or three months a year in Ireland as the chief engineer. There were many problems remaining to be solved. Jessop found that there were discrepancies in the lock sizes and he arranged for them to be standardised to take vessels 63ft 6in by 14ft. He also proposed a basin beside the Liffey. But his greatest difficulty was working out how to cross the bogs that lay between Dublin and the Shannon. The solution was to start laying out drainage ditches on either side of the proposed line of the canal in a square grid network that started at a point 55 yards from the centre line. As the ground gradually dried out, the squares left between the drains were dug up and used as spoil to build an embankment. In this scheme, it was not necessary to reach the hard ground beneath the bog so long as the bank was stable. The proposed docks were agreed and Jessop drew up plans for the docks at Ringsend with a triple entrance lock. The objective was to join the Shannon, but for that to serve any useful purpose the river had to be navigable. Jessop found that this was far from being the case, and eventually the canal company had to add river improvements to their other commitments. The Ringsend docks were opened in 1796, but there were troubles along the main canal and with the river navigation, including a major breach in the canal and problems stabilising the great embankment across Edenberry Bog. The canal finally reached the new harbour by the Shannon in 1804. It was an exceptionally difficult canal to construct, but in the end the great bogs had been conquered.

Back in mainland Britain, canal mania was reaching a peak, with seven new canal Acts passed in 1792, twenty-one in 1793 and another twelve the following year. Just as Brindley had been the man everyone wanted as chief engineer, now it was Jessop's turn to face a huge demand for his expertise with three very important canals, the Grand Junction and the Barnsley, both authorised in 1793, and the Rochdale of 1794. The most important was undoubtedly the Grand Junction, which was intended as part of a more direct route from the Thames near the centre of London to Birmingham, bypassing the rest of the river and avoiding the inconveniently winding lower part of the Oxford Canal. There were also to be several short branches so that altogether there would eventually be 124 miles of canal to construct, much of it through difficult country.

The original survey had been by James Barnes in 1792 and Jessop praised his work, but suggested several variations. The route for the main line as eventually settled would be from the Thames at Brentford to Braunston on the northern section of the Oxford Canal. The first part of the route from the Thames was comparatively straightforward, with a flight of six locks at Hanwell to lift the canal 50ft. While this was intended to be a canal that could be used by narrowboats, Jessop called for broad locks that could take barges or two narrowboats side by side, a decision that did a great deal to ensure the commercial success of the canal long after trade had ended on many others. After the first steep climb there was a steady progression at first following the line of the River Brent, with a

4.2 A pair of narrowboats working through the three locks at Soulbury on the Grand Junction (now Grand Union) Canal. (Author)

scattering of locks until the first major obstacle was reached: the Chiltern Hills, running right across the proposed line of the canal. Barnes had suggested crossing over them by locks, but Jessop had a far bolder solution. He would carve right through them, not in the conventional way by a tunnel, but by a deep open cutting at Tring.

Nothing on this scale had been attempted before. The Tring cutting would have to be dug by hand, with pickaxe and shovel, the men hacking their way through the soft chalk. There was also the problem of how to get the spoil out of the deepening gash in the hillside, with sides far too steep for a man with loaded wheelbarrow to manage. The answer was barrow runs. A series of plank walkways were constructed from the bottom of the cutting to the rim. Ropes were attached to the barrows, with the other end running over a pulley to a horse that would haul the barrow up. The man's job was to keep the barrow steady as it went up, then run back down with the barrow behind him. It might not seem too problematic on a fine, sunny day, but one can imagine just how precarious it must have been on a rainy day, with the planks covered in slimy clay and greasy chalk. There are no contemporary illustrations of the scene, but a generation later when the railway came on a nearby parallel route, we do have an accurate illustration of barrow runs at Tring (p. 180).

Emerging from the cutting, the canal begins the descent from this summit level through a flight of locks set in a sweeping curve of the canal. One obvious necessity here was to provide the summit with water. The answer was an extensive reservoir and feeder: the feeder itself was made navigable and became the Wendover Arm. The first reservoir proved inadequate, and eventually three more smaller ones were added. Today they are a much-valued wildlife reserve. The most obvious canal feature near the junction is the maintenance yard, but that was only built in the middle of the nineteenth century.

The two biggest challenges on the line were represented by two major tunnels: 2,042 yards long at Braunston and 3,056 yards long at Blisworth. When Brindley had been faced with tunnelling through Harecastle Hill, he had baulked at the idea of making it wide enough to take broad barges, but Jessop was aware that a narrow tunnel would be an unacceptable barrier to smooth movement: there was little point in making locks wide enough to take two narrowboats if boats could not pass each other when travelling in opposite directions through the tunnels. He was not, however, prepared to go that bit extra and provide towpaths, so boats would have to be legged through. There were problems at Braunston, largely because of the contractors failing to do their work efficiently and the failure of the resident engineering staff to literally keep them on the straight and narrow. The result is a distinct kink in the middle, which is very apparent today when one goes through by boat. They also encountered quicksand that had not shown up in the test borings, but they still managed to have the work completed by June 1796. Blisworth, however, was to present far more difficulties.

The contactors working at Blisworth found the ground through which they were supposed to be tunnelling difficult going so, without consulting the chief engineer, they altered the alignment to take an easier route. When the work was inspected, it was at once clear that the whole tunnel was seriously awry, the brick lining was crumbling and parts had collapsed. Jessop proposed abandoning the whole enterprise, and offered an alternative solution. He would make a deep cutting across the top of the hill, with locks at either end. To keep the short summit in water, a steam engine would be installed to pump water back up when boats locked through. At the same time, Barnes, who had carried out the initial survey, said he would build a second tunnel and would complete the whole project in three years at a cost of £48,000. The canal committee chose to go with the tunnel. It was a blow to Jessop's status in the whole construction process. However, in 1797, all work on the tunnel came to a halt as the company concentrated on completing the rest of the canal. That left a gap that, as work proceeded, needed to be filled. Now Jessop was back with a solution: the two sections of canal could be joined by a cast iron tramway along the line of the tunnel that could be used until work restarted and the tunnel was opened. His idea was accepted. The tunnel was finally opened in 1805.

Another obstacle to be overcome was the River Ouse, which ran right across the proposed line of the canal in a wide valley. Once again, there were discussions on the best way to achieve this. One suggestion was to have locks at either side of the river, which could then be crossed at the level. Apart from the problem of water supply at the upper levels, there was another difficulty: the river was known to be prone to flooding. The obvious alternative was an embankment, with the river itself crossed on an aqueduct. A conventional stone structure was completed, but was far from satisfactory and by 1807 it was reported that piers had been damaged, arches were out of alignment and part of the embankment had caved in. There was now a major argument about who should pay for the repairs: the company saying it was down to bad workmanship by the contractors; the contractors pointing out that it had been passed as properly constructed when completed, and faults were down to poor maintenance. The argument was never settled as repairs were soon put out of the question: one of those regular Ouse floods crashed down the valley and brought the whole structure down. A solution was eventually found, and a new cast iron aqueduct built that still stands. This use of iron is a subject we shall come back to later.

The Grand Junction when it finally opened was a great success and eventually had branches open to Wendover, Northampton, Buckingham, Aylesbury and Slough. The most important branch left the main line at Bull's Bridge and ran for nearly 14 miles to Paddington, then at the edge of London. This branch would later be extended via the Regent's Canal to the Thames at Limehouse.

The Barnsley Canal was designed to join the Dearne and Dove Canal that began at Swinton on the River Don and finished at the Dearne near Barnsley. Both canals were authorised at the same time. The Barnsley was to continue the line through to the Aire & Calder, passing through a country rich with coal mines. The lock dimensions were determined by the connections, as the canal had to be usable by the river barges, so they were 78ft 6in long and 14ft 6in wide. It was not a long canal, just 11½ miles, but required two major engineering works. The first was an aqueduct across the Dearne. Jessop's first thoughts were to build an embankment across the valley and cross the river in a single arch, but further investigation showed that there was a good solid foundation of rock on either side of the river, so opted instead for a five-arch aqueduct. Sadly, no traces of this grand structure remain. The other obstacle to be overcome was a high ridge across the line of the canal that Jessop decided could best be tackled by a cutting. This remains a very prominent feature in the landscape, as impressive as the more famous Tring cutting. One can still see blocks of sandstone that have been blasted away and tumbled down the hillside. Water supply was provided by the Wintersett reservoir and a second reservoir was later added right beside the canal at Cold Hendley. So far the route had remained on the level and, after a final wriggle through a hummocky landscape, it finally descended to the Aire & Calder through fifteen locks in just 3 miles. It was closed in 1953.

The first canal across the Pennines, the Leeds & Liverpool, was begun in 1770, but by the 1790s was still far from complete. It had coped with the hills by avoiding them as far as possible, taking a great sweeping curve to the north. The canal runs for 127 miles, compared with the current shortest road route of just 74 miles. The Rochdale Canal was to take a far more direct route, but would inevitably involve many locks. There were two surveys carried out by John Rennie that resulted in two applications to Parliament. Both bills were fiercely opposed by the mill owners along the route, who complained that the proposals would have left them without enough water to work their machinery. Jessop was then persuaded to take on the project, and his survey demonstrated that it would be perfectly possible to supply all the water the canal needed from reservoirs; his survey had shown that reservoirs could be built in the area as the soil was entirely suitable. This removed the main objection to the bill.

Rennie's plan had also called for a 3,000-yard tunnel at the summit. Jessop's plan did away with the tunnel altogether, but instead called for more locks to carry the canal higher up the ridge, which, like the Barnsley, would be pierced with a long cutting. Lock construction called for immense resources. The canal was to stretch from the Calder & Hebble at Sowerby Bridge to Castlefield Junction in Manchester, where it joined the Bridgewater Canal. It was just 33 miles, but it required thirty-six locks to take it from the eastern end to the summit and fifty-six from the summit to Castlefield, and they were not narrow locks, but 14ft wide and 74ft long. One of the interesting features of the canal is the bridge at March Barn that, instead of crossing the canal at a right angles, is built on the skew. This entailed a complex system of laying the stones in winding courses.

The reservoirs were the key to the whole project. Jessop originally planned for two, at Hollingsworth and Blackstone Edge. The Hollingsworth reservoir was contained behind an immense bank, 10ft wide at the top to take a roadway with a slope of 1:2 and a 9ft-thick core of puddled clay at the centre to keep it water tight. It must have been a scar on the land-scape when it was first built, but is now considered a local beauty spot and renamed Hollingsworth Lake. Later a third reservoir was added.

4.3 A bridge on the Rochdale Canal, with a trap at the bottom of a small feeder stream to stop silt entering the canal. (Author)

Even with the guarantee that water was available, Jessop did his best to avoid waste. He arranged that the locks on the Manchester side of the summit should have a uniform fall so that, in his own words: 'The water that serves one will serve the rest without waste.' In various places, small hill streams fell down to the canal, and Jessop made use of them by building stone settling tanks to trap any sediment before it reached the canal. The whole work was completed in 1804.

The other trans-Pennine canals were not faring so well: the Huddersfield Narrow, begun at the same time as the Rochdale, was only opened in 1811, while the Leeds & Liverpool finally opened in 1816, nearly half a century after work began. That the canal was a great success can be judged by the immense amount of mill and warehouse building in many places along its banks.

Samuel Smiles was full of praise for the engineer and his works: 'The skill and judgement with which he planned them reflected the greatest credit on their designer: and whoever examines the works at this day … will admit that the mark of a master's hand is unmistakably stamped upon them.' No doubt Jessop would have been pleased had Smiles not attributed the work to the wrong engineer. He attributed it all to Rennie, whose plans had actually been rejected.

Another canal project that was to occupy him at this time began with plans to connect the Dee at Chester to the Severn at Shrewsbury. The Chester Canal had already been built as far as Nantwich, but had then come to an abrupt end. The only hope for it ever making any money was for it to have useful connections, and the busy iron works of North Wales offered a potentially valuable source. Various proposals were considered, but it was Jessop who came up with the solution. The main line of the canal would stretch from a junction with the Chester near Nantwich and then head for the industrial area north of the River Dee.

As always, Jessop was meticulous in his planning, suiting the methods to be used to the terrain. It leaves the Chester Canal via a flight of four locks and now, with no major towns nor even large villages to call in on for trade, Jessop was able to take the best and easiest line, going straight where he could and meandering where necessary to avoid obstacles. Locks are well spread out, but do come together in a three-lock staircase at Grindley Brook. At the heart of the canal is the market town of Ellesmere that the canal was originally named after, although we no longer use that name, the surviving central section now being called the Llangollen Canal.

The question of water supply was different from that of the other canals that Jessop was involved with at the time as it does not cross a watershed, but rises steadily. Jessop planned to build a weir across the river near Llangollen and divert water down a navigable feeder canal through the town. One obstacle was the peat bog of Whixall Moss, but Jessop was, of course, experienced with land drainage schemes. He was able to drain the land sufficiently to reach a firm foundation on which to build up an embankment. There were two valleys to cross: the Ceiriog at Chirk and the Dee at Froncysyllte. We shall be looking at those in the next chapter. With the planning complete, Jessop had hoped to appoint an experienced canal engineer of his own choosing to supervise construction, but the proprietors insisted he take on a Scotsman who had impressed them with his abilities. His name was Thomas Telford.

Among the many calls on Jessop's expertise was a request to advise on a possible canal down the Wandle valley from Croydon to Wandsworth in south London. He advised against it, suggesting instead that the area would be better served by the construction of a railway, of the sort already common in northern England and which he himself had used as a temporary measure at Blisworth. This was agreed and the Surrey Iron

4.4 The Surrey Iron Railway. (Young & Co. Brewery)

Railway Company was formed. It was built as a double track, with basically L-shaped rails, the vertical flange keeping the wheels on track. The height of the flange had to be reduced to an inch where the line crossed normal roads. As with other lines of the period, the rails were mounted on stone blocks, leaving the space between the rails free for the horses that would do the work. It was London's first railway, but few traces now remain, apart from some of the stone sleeper blocks with the characteristic hole in the centre into which a wooden plug was fitted to take the spike to hold the rail. When the line was closed, some were used in the wall of Young's Brewery in Wandsworth.

London was also the scene of some of Jessop's biggest and most important projects, as he was employed from 1796 to 1799 by the City Corporation as consulting engineer to consider improvements in the Port of London. The need was obvious. Traffic had been growing throughout the eighteenth century and ships were still being laid up in the Pool of London, where they moored and their cargos were shifted between ship and land by lighters. Plans were drawn up by Jessop and the City surveyor George Dance for a new complex of docks and a canal. Dance is an interesting man, a Royal Academician as well as a surveyor – and one of the few portraits of Jessop is a drawing by him. Various alternatives were

4.5 The City Canal, London. (Museum of London PLA archive)

discussed and the scheme eventually accepted was for the West India docks and the City Canal, both situated on the Isle of Dogs. There were to be two parallel docks, one for export, one for import, joined to the river by basins and locks at Limehouse and Blackwall. The canal was cut across the Isle alongside the docks and was designed to take the biggest ships of the day, with a width at water level of 176ft and a depth of 23ft. Access to the tidal river was through locks at either end. There was one major problem to overcome in constructing the canal. The land was 6ft below the highest tide level, so the canal had to be run on an embankment, constructed of excavated gravel with a puddled clay core. There was one mishap just before the canal was due to open, when an exceptionally high tide caused damage at Blackwall, but that was quickly repaired and on 9 December 1805 the first vessel, the *Duchess of York*, was towed through the canal.

Work began on the West India Docks in February 1800, starting with the import dock, 2,600ft long and 510ft wide, with the main ship entrance lock at Blackwall being 193ft by 45ft. Six five-storey warehouses were built along one side. In 1803, work started on the smaller export dock, the same length as the export, but just 400ft wide. It was formally opened in 1806. The docks were at the time the largest of their kind ever built, and for the first time ships on the Thames could unload at the quayside at all states of the tide.

Bristol was, in the eighteenth century, Britain's second busiest port, but was beginning to face increasing competition from Liverpool. It suffered from the same problem as London, in that vessels were reliant on the tidal river and could only reach the quays when conditions were right. Smeaton had looked at the problem, but his proposal was not followed up. Various other proposals were considered over the years, but it was not until 1802 that a final plan put forward by Jessop was approved. The idea was to create a floating harbour by building an entrance lock and basin to provide access, while diverting the river itself into a new cutting. The old quays along the River Frome would be preserved and there was to be a small basin, Bathurst Basin, near the eastern end. A canal and feeder would run from Temple Meads to rejoin the natural river above Netham Weir. When work began in 1803, Jessop's son, Josias, was appointed resident engineer. The whole scheme was completed in 1809. It ensured the Port of Bristol remained viable until the middle of the twentieth century, when the new deep-water port of Avonmouth was opened.

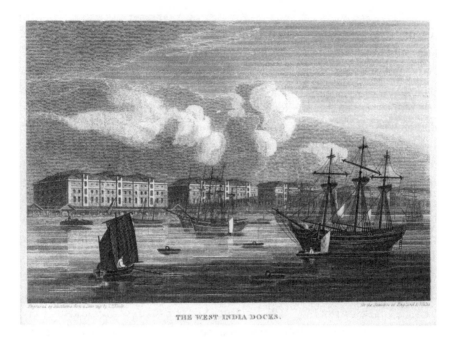

THE WEST INDIA DOCKS.

4.6 The West India Docks on the Isle of Dogs, London. (Museum of London PLA archive)

Jessop was not only in charge of some of the most important engineering projects of the age, but was also consulted on many, many more and also played an important part in the development of the Caledonian Canal. Jessop died at Butterley in November 1814 at the age of 69. Where Brindley had been the man everyone wanted to have for their own canal schemes, in the mania years of the 1790s that role was taken by William Jessop.

5

THOMAS TELFORD

Thomas Telford was born on 9 August 1757 in a cottage in Eskdale in the Scottish Lowlands, where his father was a shepherd. It was a happy event for the family, as their first son, also Thomas, had died in infancy. But the happiness was short lived, for his father died just two months later. This was a double blow, for the family lived in a tied cottage and they had to move out to make way for the new shepherd. They had to take a single room and were forced to rely on mother Janet's brother for charity. The boy received a rudimentary education, and it was soon time for him to begin earning a living. He was first apprenticed to a stonemason at Lochmaben, but was so badly treated he was soon back home. Fortunately a relation of Janet Telford, Thomas Jackson, managed to get him another apprenticeship with a Langholm mason, Andrew Thomson. We have two reminders of his time there. He was involved in the construction of a bridge across the Esk and, although it has been widened, Telford's mason mark can still be seen. On a more personal note, he carved a headstone for his father and the young Thomas he had never

Image: Thomas Telford. (From Smiles, *Lives of the Engineers*, 1862)

known. It can be seen in the churchyard at Bentpath, and the lettering is neat and precise, a tribute to his craftsmanship. One other event had a marked effect on his life. He was befriended by a wealthy lady, Elizabeth Pasley, who had a passion for education. Young Thomas was an eager pupil and, thanks to her, he developed a deep love of poetry. He was to write verse himself and was later to become a close friend of the Poet Laureate Robert Southey.

Telford was ambitious, and was aware that there were few opportunities for advancing his career if he stayed in Eskdale. So, in 1780 he set off to walk to Edinburgh. His timing was ideal, for the city was being transformed by the construction of the fashionable New Town. He not only found work, but began to learn about the latest architectural styles. He also learned what he described as 'the Art of delineating Architecture upon Paper'. He returned briefly to Eskdale at the end of 1781, but by January 1782 he felt sufficiently confident to make his way to London to try and develop his career.

He was fortunate in receiving help from the Pasley family. John Pasley introduced him to the leading architects of the day, Sir William Chambers and Robert Adam, a connection that helped him get work on the construction of Somerset House, designed by Chambers. Then another opportunity opened up for him when Sir James Johnstone decided he wanted to make improvements to his grand house, Westerhall, near Telford's old home in Scotland. Sir James wrote to his brother William in London, who had married into the wealthy Pulteney family – a name still known through the famous Pulteney Bridge in Bath. Pulteney was delighted to find a mason who was educated, understood architectural drawings and had his own views on architecture. These connections with the wealthy and influential of the land undoubtedly helped further Telford's career, and his next move was to Portsmouth, where major works were in hand to improve the naval dockyard. He was no longer just another mason on the project, but a supervisor, one of whose tasks was to oversee the construction of a chapel and the grand Commissioner's House. He was to spend two years at Portsmouth, and in what little spare time he had he spent in studying various subjects including chemistry, but with a very practical rather than merely theoretical interest. He paid to attend lectures on the chemistry of mortar and quicklime, essential elements in any building project.

Once again, the Pulteney connection brought him an interesting job: Sir William was elected MP for Shrewsbury and he decided that, with his new status, he should renovate part of his estate, Shrewsbury Castle. Telford was given the job, and in 1786 he set off for Shropshire in what was to prove a pivotal moment in his career. He thoroughly enjoyed his time there and, as well as being responsible for the castle, he was appointed County Surveyor. This was one of those posts that could mean a great deal or very little: some counties went in for public works on a grand scale, others did very little. Fortunately for Telford, Shropshire proved to be one of the former. This gave him an opportunity to widen the scope of his work. One of the less obvious tasks he undertook was to oversee the excavation of the Roman remains at Wroxeter. He also had the opportunity to indulge his passion for architecture, designing two churches. The one at Madeley is frankly rather dull, but the other, St Mary Magdalene at Bridgnorth, is altogether more appealing, occupying a key position in the town. It is built in the then fashionable neo-classical style, with a tall, domed clock tower.

More importantly, given his future career, Telford was soon involved in road construction and bridge building. His first bridge over the Severn was a conventional stone structure, crossing the river on three sandstone arches. But it seems he was at this stage more interested in carving out a career as an architect than as a civil engineer. With that in mind he took a trip to Bath to see the new buildings of the city that epitomised all that was best in Georgian architecture. But then, in 1793 he was able to write to friends in Scotland that he had been appointed 'Sole Agent, Architect and Engineer' on the Ellesmere Canal, thanks to the lobbying of the Shropshire supporters and in spite of Jessop wishing to appoint one of his own assistants, William Turner. Whatever Jessop's misgiving may have been, Telford was to prove more than able to superintend work on the canal. He still retained his post as County Surveyor.

Telford brought down an old friend and associate, Matthew Davidson, to work as his assistant, and things went along smoothly until, after a little more than a year on the job, he found himself with a new position as chief engineer for the Shrewsbury Canal. It had been begun under the direction of Josiah Clowes, but he had died suddenly before the work was completed. It was not a major work, just 18 miles long, and had just one big engineering feature, an aqueduct over the River Tern at Longdon.

The river was not navigable, so Clowes had begun work on a conventional masonry aqueduct. But, while still under construction, a sudden flood had demolished the partially built arches. The obvious option was simply to start again and rebuild them, but the chief promoter of the canal was the local iron master William Reynolds, who encouraged Telford to try a different approach: to build an iron aqueduct. The abutments were cut back and faced with stone, leaving a 187ft gap to be bridged. The iron trough consisted of twenty-six sections bolted together with the towpath slung on the outside, the whole structure supported on iron struts. The great advantage of the aqueduct was that the piers had a comparatively light structure to support, just iron plates instead of heavy masonry lined with puddled clay.

When he returned to the Ellesmere Canal, Telford felt he had the answer to the problematic crossing of the Dee valley. Before he could tackle that problem, however, he had another to deal with in his role as County Surveyor. The floods that had destroyed the aqueduct at Longdon had also swept away the bridge over the Severn at Buildwas. Telford

5.1 Telford's innovative iron bridge at Buildwas: it no longer exists. (Shropshire Records and Research Service)

once again turned to iron. He was familiar with the original bridge at Ironbridge. The very high arch allowed the stone abutments to move inwards. He planned to use a much flatter arch with a span of 130ft and a rise of just 24ft. It was a success, although the bridge no longer remains. He could now return to the Dee crossing.

The original proposal had been to reduce the height of the aqueduct by 50ft by means of locks on either side of the valley, which at the same time would reduce the length of the span. There would have had to be a pump back system or the aqueduct would simply have acted as a sump. Telford now sent in his suggestion for an iron trough aqueduct to carry the canal at the full height above the river of 125ft, and it was to be carried on eighteen spans, supported by masonry pillars that would be solid at the bottom, but hollow and cross braced in the upper parts. The towpath would be inside the trough, cantilevered out over the water. It was the most remarkable structure of its time and still impresses to this day. There is, however, a certain controversy surrounding it.

5.2 Pontcysyllte aqueduct carrying the Ellesmere Canal high above the River Dee. (Author)

Most accounts refer to the structure as Telford's masterpiece, yet conventionally the credit goes to the canal's chief engineer, and that was still William Jessop. He would have had the final decision on whether or not it should be built and by what method. Charles Hadfield, the doyen of canal historians, argued that Telford had unreasonably claimed all the credit for himself. He and I argued amicably about this over many years, though we agreed that both men should take a share of the credit. Contemporaries certainly lauded the work as Telford's, and I am inclined to take a charitable view of what happened. Jessop was a modest man, and he was quite prepared to let his younger colleague take the praise, knowing no doubt that it would help make his reputation in the world, which indeed it did.

An iron trough was also proposed for the other big aqueduct at Chirk, but the owner of nearby Chirk Castle objected to it and wanted something altogether more picturesque. As a result, Chirk appears to be a conventional, handsome masonry aqueduct and few who cross it are aware that the water is actually held in an iron trough, embedded in the stonework.

As work continued on the Ellesmere Canal, Telford had a number of other projects to keep him very busy. He was involved in work for the Liverpool Water Works, a grand name for a less imposing concern. He was, however, able to improve the water supply for the local community by a piped system using a quite small Boulton & Watt steam engine, with an 18in-diameter cylinder to work 12in pumps. If one compares this with the Boulton & Watt engine installed not to supply a whole community, but simply to replenish the Kennet & Avon Canal at Crofton, where a 42in engine drives 30in pumps, one can see its limitations.

A far greater work attracted his attention. The medieval bridge across the Thames in the heart of London had long been a major obstacle, with its narrow arches making passing through it dangerous and sometimes impossible. It obviously needed replacement. Telford's design called for a cast iron structure that would soar across the whole river in a single 600ft span. According to his calculations, it would require 6,500 tons of metal and to resist the pressure from the arch would require abutments built of 12,000 cubic metres of granite blocks. Considering that his previous bridge at Buildwas had only needed 173 tons of iron, one can see that this was a giant leap of imagination as well as a great leap across space. The design was not accepted, which is just as well for Telford's reputation as modern engineers believe it would have failed: the abutments would not have contained the thrust. He was soon, however, to enter a period of intense activity that would take him back to his native Scotland.

Following the two unsuccessful Jacobite rebellions of 1715 and 1745, the Scottish Highlands were in a depressed condition, made even worse by the Highland Clearances. The latter involved landlords evicting crofters from their smallholdings to make way for sheep runs. One result was a widespread depopulation as many Highlanders simply left the country altogether, many heading for new homes in Canada. There were two problems to be solved: the lack of employment and poor communications. A network of military roads had been developed under General Wade, but in general they were neither very well constructed nor very sensibly planned. The government felt the need to act to prevent a further wave of emigration and in 1801 Nicholas Vansittart, Joint Secretary to the Treasury, wrote to Telford asking him to investigate ways in which the fishing industry of Scotland could be improved and at the same time to look at ways in which the harbours of the west coast might be improved for use by both commercial craft and the navy. He was told that more detailed instructions would follow, but Telford was so enthusiastic that he was not prepared to wait. He set off for a whirlwind tour of Scotland, at the end of which he put forward a new proposal that would he was sure improve communications and offer employment for many Highlanders – a canal across the country from Fort William to Inverness.

This was not a new idea: James Watt had earlier made a similar suggestion, a fact that Telford mentioned in his own report. It was an obvious route that followed a natural fault line down the Great Glen. An artificial waterway could be built to link the sea via Loch Linnhe on the west coast to the Beauly Firth on the east, via Lochs Lochy, Oich and Ness. Vansittart responded favourably to the suggestion and asked Telford to write a full report on the proposed canal, to investigate other important transport routes, including one from Carlisle to Portpatrick, and also to report on the state of emigration from the Highlands. The latter requirements suggest that the government saw the canal as providing essential work as well as a valuable new transport route that, in particular, would enable vessels to move easily between the Irish Sea and the North Sea. In 1803, the Act was passed authorising the construction of what was now known as the Caledonian Canal.

The organisation for developing the project was a little complex. Telford certainly seems to have been the chief engineer, with two resident engineers working under him. The secretary was John Rickman and the land agent responsible for negotiating the rights to the land was James Hope. The governing body, however, felt that it was their duty, given the importance of the scheme, to get a second opinion: 'We took measures as early as the 4th of August, for obtaining the opinion and assistance of Mr William Jessop, another eminent and experienced Engineer.' So, what precisely was Jessop's role? It is hard to say. As far as we can tell he was a consultant engineer to the project, and there is no evidence of any clash with Telford as the work progressed. We do know one instance when Telford bowed to his old boss's judgement. Telford had been very concerned to keep costs down to a minimum, and had suggested using turf-sided locks. Jessop persuaded him that these would only provide endless trouble in the future, and that stone construction would be far superior.

Vessels from the Irish Sea approach the tidal lock at Corpach at the end of Loch Linnhe, a spot marked by a small lighthouse. In the next stretch of artificial cutting, the canal has to be lifted by 60ft to reach Loch Lochy. Instead of a gentle rise, this is achieved through eight interconnected locks, known as Neptune's Staircase. Along the way are three short aqueducts, scarcely noticeable, before the Gairlochy locks give access to Loch Lochy. Although this appears entirely natural, there was much engineering work involved, diverting the river from its original course and damming it to raise the water level in the loch. From the far side of the loch, there is a short section leading to Loch Oich. It may be short, but it was a considerable challenge, as the canal had to be driven through the deep Laggan cutting, hacked and blasted out of solid rock. Loch Oich had to be dredged. The next canalised section included a five-lock staircase at Fort Augustus to bring the canal down to Loch Ness. The very final section was to present its own problem. The Beauly Firth was too shallow for vessels to reach Inverness, so a great embankment had to be built out to sea, consolidated and then the canal had to be cut into it and a sea lock built at the end.

We are fortunate that we have an eyewitness account of what work was like on the canal. Telford took Robert Southey on a tour of sites and Southey wrote detailed descriptions of what he saw. Neptune's Staircase

5.3 Neptune's Staircase on the Caledonian Canal. The poles stuck up in the air were used with capstans to open the lock gates.

impressed him hugely, but work there had already been completed. At Laggan cutting, he described the barrow runs, operated by horses, where the spoil wagons were hauled up from the cut and connected via a pulley to the empty wagons that were lowered down a parallel plane. One amusing fact he noted was that the rest time for men and horses was announced by blowing a horn but, if it was late, the horses simply stopped anyway. He saw the lock construction at Fort Augustus, where three steam engines were at work, pumping out water: the largest shifting 8,000 gallons a minute. Iron lock gates had arrived by sea, and a steam dredger was at work at the entrance to the lock. The men who worked there were almost all Highlanders and, although they were not paid as much as professional navvies further south, they were far better off than other labourers in the region. The canal did indeed provide welcome relief for many. It did not, however, live up to its promoters' expectations. It was hoped to provide a useful route for naval vessels, but they never used it.

The canal was by far the biggest of Telford's works in Scotland, but by no means the only one, nor necessarily the most beneficial in the long term. He was also involved in building roads, harbours and bridges throughout the region. Road construction in Britain had not improved since Metcalf's time, but Telford set out to devise the best system for roads that would provide a good, hard surface that would survive bad weather. He described the process in detail for a road being constructed on an embankment. Once the ground had been consolidated, gravel had to be laid on top with a curved surface, 4in deep at the side and 8in in the middle. All the gravel had to be screened and any stone longer than 2in removed. Once the gravel had been laid, carts were wheeled over and any hollows filled in. The next layer was similarly curved, but 6in in the middle and 2in at the side. The stones had to be shaped as nearly cubical as possible and had to be able to pass through a 2½in-diameter ring. The roads were certainly good and solid, but very expensive to construct.

Obviously, Telford was unable to personally supervise such a wide-ranging programme of road construction throughout the Highlands. The area was broken into four divisions, each with its own resident engineer, who would be in charge of the works. Telford certainly laid down the lines of the routes, and left his instructions on the processes to be followed. He was a regular visitor to Scotland to see the progress for himself. When it came to major river crossings, he was responsible for designing the bridges. His first was across the Dee at Tongland, a simple stone bridge crossing the river in a single segmented arch of 110ft span. His next two, however, were iron. At Bonar, the crossing of the Dornoch Firth was intended to consist of two symmetrical arches, but when tests were made Telford discovered that there was only a solid rock bottom on side of the river, so the middle pier had to be resited, resulting in one arch of 60ft and the other 50ft. The bridge withstood a battering in a flood of 1814, when pine logs crashed against it, and again in 1818, when a schooner ended up running into it. The schooner lost its masts, but the bridge stood firm. It did not, however, survive severe floods in 1892. The other iron bridge presented no such problems. It crosses the Spey at Craigellachie in a single graceful span, with the ends marked by crenelated towers. A plaque on one of the towers announces that it was opened in 1814 and that the iron work had been supplied by Plas Kynaston, the company responsible for the trough on Pontcysyllte.

The first major coastal project with which Telford was concerned at the time was for developing Wick as a safe harbour for a fishing fleet. It was a natural site for a harbour, based on a sheltered bay – the name comes from the Viking 'vic' for bay. It was to be developed as a new town, originally given the name 'Pulteneytown'. The old town was all on the north bank of the river, the new on the south. It was handsomely built, divided into two distinct areas. The lower town was based on the harbour and contained a curing station in the inner harbour, where herring were prepared, salted and packed into barrels. Living quarters were built into the same structure, which must have been decidedly aromatic. There was also a smoke house in the town. The upper town was a far grander residential area, laid out on a regular grid and, if not quite as elegant as the other town associated with the Pulteney name, Bath, is still very handsome. The improved port was a huge success and Sir John Sinclair, the prime mover behind the scheme, was able to report that as many as 100 vessels were often to be seen in the harbour and that the annual catch amounted to 200,000 barrels of herring. His next task took him to Aberdeen, where he was to work on improvements to the harbour, work that had been started by Smeaton. The main job was to lengthen the north pier and construct a breakwater.

In the middle of all his busy life, working in Scotland, Telford received a letter from Count Baltzar von Platen, who had recently retired from his post as head of the Swedish Navy. He wanted Telford to be the engineer for a project of his own: a ship canal that would cut across from the North Sea to the Baltic. A start had already been made with the construction of the Trollhätte Canal that linked the navigable River Göta to Lake Vänern. The engineer in charge for that project had been a Norwegian, Samuel Bagge. It was felt that there was not enough experience for the project to be carried through without the help of a leading engineer, and King Gustav IV had personally approved the works and suggested that an 'experienced Englishman' should be called in – a description that might not altogether have pleased the proud Scot. There is a suggestion that Bagge had already been to Scotland and seen the works on the Caledonian, and had approved the request to Telford.

There was a problem for Telford. He would need to visit Sweden, but Britain was engaged in the Napoleonic wars and he could only go if a suitable armed naval vessel was available. It was arranged for Telford to join a British convoy leaving Leith. The list of provisions provided for Telford during the voyage now hangs by the entrance to the restaurant at the Institution of Civil Engineers. He was clearly no teetotaller, for the list contains two dozen bottles of Madeira, two dozen of Port, three dozen of cider, three dozen of Piny, six dozen of porter, and half a dozen bottles of gin and the same of brandy. When he arrived, Telford had the advantage of an earlier survey of the proposed route, using a string of lakes joined by 53 miles of artificial waterway. The most difficult section involved the link from Lake Vänern to Lake Vättern, which would require a flight of fifteen locks, eight arranged as connecting pairs and the other seven in a staircase. Much of the land involved cutting through hard granite, so Telford decided that it was impractical to make it the same dimensions as the Caledonian. Instead of being able to take vessels 150ft by 35ft, the Göta Canal was limited to vessels 103ft by 23ft. Telford spent a total of six weeks going over the ground and two weeks staying at von Platen's house drawing up the plans, before returning to Britain.

By 1809, work had been authorised and Bagge returned from Norway to oversee the project. The works were being paid for by the state and at the start there were around 900 soldiers drafted in, together with 200 labourers and 150 Russian prisoners of war who had decided to stay in Sweden. What was lacking was basic expertise and equipment. Telford sent out detailed drawings of locks and bridges and, visiting the canal today, one can see many similarities with structures on the Caledonian. He also sent over examples of the equipment used in Britain: three railway wagons, six sections of rails, three wheelbarrows, three picks and three shovels. In 1813, after Napoleon's army had been forced out of Russia, Telford was able to return and found the work to be well done. Two young Swedish engineers returned with him for training in Scotland for nine months.

Things seemed to be going well when tragic news reached Telford. Bragge had been drowned, his boat caught in a storm on Lake Vättern. Telford had left two of his own engineers behind and they were to take responsibility, but one of them, James Simpson, proved to be a troublemaker and was sacked. That was not the only problem that needed addressing. Telford had sent von Platen plans for both timber and iron lock gates and iron bridges. The problem was that they were cast iron; Sweden had a thriving iron industry, but it only produced wrought iron. It would have been possible to have had everything made in Britain and shipped over, but instead James Thomson, who was an associate of Hazeldine of the Plas Kynaston foundry, went to Sweden to establish an iron works at Motala, where the canal joins Lake Vättern. It was entirely successful and helped to establish the town as an important industrial centre. The canal was officially opened in 1830 when the royal yacht made a journey from coast to coast. Von Platen was not there to see it – he had died the previous year – but his monument still stands beside the canal he had done so much to promote. Telford had already been honoured for his part in the project, having been made a Knight of the Royal Order of Vasa, after which all letters from Sweden were addressed to Sir Thomas Telford.

5.4 The opening of the Göta Canal in September 1832, painted by Johan Christian Berger. (The Göta Canal Co.)

In 1815, the war with Napoleon ended at Waterloo and, although peace was welcomed, there was widespread poverty throughout the country, with many on low wages or unemployed. The government reacted by pushing through the Exchequer Bill Loan in 1817, which provided £1,750,000 for public works; Telford was appointed technical adviser for civil engineering. His main task was to build a new road from Shrewsbury to Holyhead for the mail coaches carrying the Irish mail. The route is roughly that of the modern A5. He had already been working on improvements to the route. One of the most dramatic sections is the road through Snowdonia down the Nant Ffrancon pass. The section gets a fine flourish at the start with the bridge over the Conwy at Betws-y-Coed. It is an elegant cast iron arch, with decorations in the spandrils of rose, thistle, shamrock and leek to represent the Act of Union. It also carries the message 'this arch was constructed in the year the battle of Waterloo was fought', although it only opened for traffic a year later. As the road was built through the mountains, immense engineering works were required, although they are not immediately obvious when using the road. However, viewed from a short distance, immense retaining walls can be seen, particularly obvious on the approach to Llyn Ogwen. The road alignment has never been changed, simply because it fills the whole space between the lake and the rocky slopes of Tryfan. In other places, the works were not so dramatic, but improvements were essential. In one of Telford's reports he described how, on the road to the west of Llangollen, deep ruts had been cut in the centre of the road by coaches, with the result that when two coaches met head on they had to stop and fill in the ruts to get out of them and pass each other. He set about converting the old narrow road of poor surfaces and sharp bends to a modern broad highway.

Although Telford's first task had been improvements in Wales, it was soon obvious that there was little point in improving the western end of the route if the rest of the system was still in a bad condition. In 1818 he was given the tedious job of having to survey the entire route from London to Chirk. He discovered sad inadequacies that had to be remedied, but they could all be dealt with using the methods already in use on his other roads. There was, however, still one major problem to overcome: how to cross the Menai Straits, separating mainland Wales from Anglesey.

Telford's original plan was for a single arch of cast iron, with a 500ft span, that had to be high enough to allow tall-masted ships to pass underneath. On either side there would be smaller, stone arches to carry the approach roads. There was an obvious problem to deal with: how to assemble the individual components without being able to use any form of scaffolding. Telford's proposed solution was to erect towers on the abutments with rollers on top. Cables would pass over the rollers so that the ironwork could be inched across. The plan was never put into operation, but it seems doubtful that the proposed arch would have withstood the high winds of the Straits.

There was, however, another solution that Telford had already investigated when contemplating a bridge across the Mersey at Runcorn: a suspension bridge. Telford was not the only engineer thinking about this form of construction. In America, James Finlay had built suspension bridges using iron chains as early as 1800. In Britain, Captain Samuel Brown, a former naval officer, had begun in a slightly different way, by first investigating the use of chains for anchors. As very little was known about the strength of chains, he built his own testing machine in 1813. Telford got in touch with him, and the two men had several discussions on chains and bridge building. Brown patented a wrought iron flat link for chains in 1817 that he used on two bridges. One at Berwick was short lived, blown away in a gale after just six months in use, while the Union Bridge across the Tweed, completed in 1820, still survives.

Telford was to use the flat link chains when he decided to abandon his plans for an iron bridge at Menai in favour of a suspension bridge. Approval for the bridge was given in 1817, in spite of some local opposition, including from the Marquis of Anglesey, who complained it would disturb his privacy. The following year, the plans were complete and work could begin. A quarry was opened up and a barge and a schooner acquired to bring materials to the site, though the former was nearly wrecked during a storm in the Strait. The first stage was the erection of the two pyramidal stone towers to hold the chains. Rocks had to be levelled on the Anglesey side, and an approach road on stone arches built at both ends. This left a central span of 579ft rising 100ft above the water. As work went forward on the towers, the iron work was being prepared, with each 9ft-long link being tested to a 100 per cent safety margin, heated and dipped in oil to prevent corrosion and dried in stoves. There was still uncertainty over factors such as how the links would respond to the stresses and loads, so an experimental rig was built near the bridge. The chains were attached to a solid support at one end and then fastened to a crank at the other and passed over shear legs, with different weights then added. There was then the problem of working out the different lengths of rods that would be needed to attach the chains to the road platform. One might think this was a simple exercise in mathematics and, indeed, in the official book of the construction printed shortly after the bridge opened, equations are given, but they were never used in practice. Instead, a scale model was built. There was also the need to anchor the chains securely, and this was done on the Anglesey side by burying the ends in a tunnel cut deep in the rock.

In 1825, everything was ready to set the chains in place. They had been built in two sections and passed over roller sets on the tops of the towers. The two ends of the chain were connected on a raft in the centre of the waterway. After that, teams of men were used to haul the chains to the right curvature. Two brave workmen celebrated the event by walking across the chains from one bank to the other. It was now possible to start adding the rods, and finally the deck itself could be attached. Everything seemed to have gone smoothly, but no one had allowed for

5.5 The suspension bridge across the Menai Straits. (Author's collection)

the wind factor. When gales blew, there was a tendency for the decking to twist and further strengthening ties had to be added. Once that was done, the problem was solved. The bridge was officially opened on 30 January 1826. It can be considered Telford's masterpiece, but he certainly felt that he had taken suspension bridges to their limit. A second suspension bridge was built at Conwy, but presented a far lesser problem. As well as the major engineering works, Telford also designed toll houses along the road, one of which at Holyhead is now a café.

Throughout the early years of the nineteenth century, Telford was constantly being called in to advise on a whole variety of road and canal schemes throughout the country. One of the more important was the Gloucester & Berkeley Canal, a ship canal, now known as the Gloucester & Sharpness. Although only 16 miles long, work that began in 1793 still languished when Telford was called in to advise in 1822. One of the problems stemmed from hiring a whole bevy of small contractors, a system insisted on by the committee. Now that work was stopped, Telford was able to hand the whole job to one major contractor, Hugh McIntosh. The project had been costing far more than estimated, and one expensive item appeared to be the stone ordered by the resident engineer, John Woodhouse. That might have been forgiven if he had not been ordering it from his own son. He was soon sent packing. The canal was completed in 1827.

Not every scheme in which Telford was involved came to fruition. He was consulted on a proposed Liverpool Ship Canal, but suspected corruption and backed away. The proposed English & Bristol Ship Canal fared better, in that an Act was passed, but work never started, which is probably just as well as it would probably never have paid its way. Telford was also consulted on various new road schemes, such as the stretch between Shaftesbury and Honiton, which he found hilly, winding and narrow, and was forced to point out that improving it would be very expensive. No action was taken. He was, however, once again involved in bridge building across the Severn. Two were handsome iron bridges, one at Holt Fleet near Worcester and the other at Mythe Bridge near Tewksbury. Ideally, the third Seven bridge, at Over just north of Gloucester, would have been of iron as well, but the local authorities thought it too undignified for a cathedral city and demanded a stone structure, which was duly built. When the present A40 was modernised, the old bridge proved too narrow and now stands rather forlornly just south of the dual carriageway.

The most important new project of this period took him to London. Although Jessop had provided the capital with a splendid new dock complex, it had open wharves and pilfering was always a problem. It was decided to build an enclosed dock near the Tower of London. Telford was in charge of the actual dock construction and the architect who would design the surrounding warehouses was Philip Hardwicke. To save space, the upper storeys of the warehouses were to overhang the wharves, supported by cast iron pillars. The Act approving construction allowed for compulsory purchase of the land, which would involve demolishing around a thousand homes and St Katharine's Hospital. The latter gave the area its new name: St Katharine's Dock. The 180ft-long entrance lock could be filled in just over five minutes by using water from the basin combined with river water pumped by an 80hp steam engine. While the lock was being filled, a second smaller engine pumped water up into the basin from the river to prevent the levels from falling. The dock opened in October 1828.

By the 1820s, there was an alternative transport system being developed that appeared to threaten the position of waterways as the country's most important transport system. A number of collieries in the northeast of England were developing steam locomotives to work on their older rail systems instead of horses. They were all private concerns, but in 1825 the Stockton & Darlington Railway was opened, the first to be approved by Act of Parliament to use locomotives. Telford was not impressed. Throughout his life, he insisted that the only value a railway would have would be to bring goods to navigable waterways. Then the far more important line to link Liverpool and Manchester was proposed and George Stephenson was appointed chief engineer. When work got under way, the railway committee felt that things were not going well and called on Telford for an opinion. He was too busy to visit the site, but sent one of his best men, James Mills, in November 1828. When Mills reported his findings, Telford was scathing and denounced the work as being mishandled. His main criticism, however, was that Stephenson had totally failed to understand that he would never be able to take his line across Chat Moss. Did the criticisms really come down to the fact that Stephenson was not doing things in the way that Telford would have done? In part that was certainly the case, and Chat Moss really must have seemed impossible to cross. But Telford was wrong. He would never get involved any steam railway projects, but continued to work on canals.

The Trent & Mersey Canal had proved a great success commercially, but the increased traffic meant that the old Harecastle tunnel, only wide enough for a single boat, was proving a serious obstacle to trade. Telford was called in and found the tunnel badly in need of repair and, even if it was brought back to a good condition, it was still inadequate. In 1824, the company finally agreed to take drastic measures. A new tunnel would have to be built alongside the old, which would be retained. This would enable a one-way system to be adopted for each of them. Where Brindley's tunnel had been low, narrow and decidedly crooked, Telford's new tunnel was straight and wide enough to include a towpath. And it was a mark of how far canal technology had advanced that, while the original had taken eleven years to complete, the new was finished in three. This was not the only Brindley scheme that was proving to present problems. His original Birmingham Canal had wandered across the land in extravagant curves and had a short summit level, with three locks to get up to it and another three to come down again. There had been improvements over the years, but it was still not able to cope with the growing traffic. Telford's proposal was bold. He would construct a canal that would run straight from the top of the Farmer's Bridge locks in the south to Wolverhampton, where it

5.6 Galton Bridge crossing the new straight line of the Birmingham Canal. (British Waterways)

would fall through another flight to join the Staffs & Worcester. It would be broad, with towpaths to either side, run on the level the whole way, and would go through the summit in a deep cutting. The old canal could not be abandoned, simply because too many important businesses had been set up along its banks, but the old bends would now be loops running off the new main line.

Because the canal was a uniform width, Telford could use a new technology to provide standardised iron bridges. These were to be cast in sections by the Horseley iron works at Tipton, and consisted of the two arches, each with a built-in handrail that could be joined by a central locking plate, with the actual footway consisting of iron plates. The biggest engineering challenge was the deep cutting at Smethwick, and this was crossed by the high Galton Bridge that used the same basic design he had used at Mythe Bridge. There was also an aqueduct to connect to the steam pumping engine on what became known as the Engine Arm. Classicism had now given way to a taste for the Gothic and the towpath is carried on a series of pointed arches.

Telford supported new canal projects, including the promotion of what would become the Macclesfield Canal. There would be one last major canal development for which he would be chief engineer – the Birmingham & Liverpool Junction, authorised in 1826. It was to start on the Staffs & Worcester, near the foot of the Wolverhampton flight, and then join the Chester Canal at Norwich. To complete the link to the Mersey there would be a short extension north from Chester. This was to be a thoroughly modern canal, taking as direct a line as possible, using cut and fill, deep cuttings through high land, and high banks over the valleys. There were to be twenty-nine locks in all, all but four grouped together in flights at Tyrley, Adderley and Audlem. It was an ambitious scheme and troubles soon appeared. The biggest headaches were the deep cuts and high banks, which proved difficult to stabilise, and none caused more problems than the embankment at Shelmore. All kinds of different approaches were tried, but none seemed to work. When Telford was too old to take active control of the work, he brought in William Cubitt to take over. He had little more success: it was only completed in 1835, and Telford did not live to see it.

Telford did not have much time for a private life and he did not socialise with contemporaries at the Smithsonians' meetings. It seems to be in

5.7 A narrowboat in one of the deep cuttings of the Birmingham & Liverpool Junction Canal – now known as the Shropshire Union.

part because he had had serious disagreements with John Rennie, who was a member, though quite what the cause was is uncertain. However, a group of younger engineers met in 1818 and decided that they needed some sort of organisation of their own, where they could meet to discuss ideas and projects. The Smithsonians was far too select a group to invite any but the well-established. The new group decided to limit membership to engineers between the ages of 20 and 35. However, it did not develop as rapidly as they had hoped, and they decided to widen the scope to create an organisation open to all qualified engineers. To give it gravitas, they needed someone to join who was universally accepted as a leader of the profession. The age ruling was hastily scrapped and Telford was invited to join. He was to become the first president of the new Institution of Civil Engineers. It achieved its founders' ambition and has become the principal organisation for civil engineers, empowered to award qualifications.

There were to be a few last bridge projects. One was for a bridge across the Water of Leith in Edinburgh and the other for a bridge over the Clyde at Broomielaw in Glasgow. Both were conventional stone structures. Telford was then consulted about a proposed bridge across the Avon Gorge at Clifton on the outskirts of Bristol. A competition had been held and he was to judge the entries. He turned them all down, including a proposal from a young Isambard Kingdom Brunel. He had already decided, from his own experience, that the span at Menai had already reached a limit beyond which it was unsafe to go. The Brunel proposal far exceeded this limit. As a result, the committee invited Telford to present a design of his own. He duly obliged, but came up with a grotesque solution. To reduce the span of the suspension bridge to 600ft, he called for two towers to rise up from the river bank, and they were not plain but elaborately Gothic. The committee accepted, but public opinion was, not surprisingly, strongly against them. They were never built, and Brunel got the job that was to mark the turning point in his career.

At the age of 75, Telford decided to give up public life, but when the Duke of Wellington made a personal request for him to visit Dover to try and solve the problem of silting in the harbour, he felt unable to refuse. He visited the site in January 1834 and wrote what proved to be his last report. He died in September of that year. He had asked to be buried in the parish church of Westminster near his London home, but the Institution of Civil Engineers had other ideas. Instead of the humble church, he was buried in the great Abbey. It was perhaps an appropriate place for a public-spirited man who had always loved fine architecture.

6

JAMES WATT

James Watt was born at Greenock on the Firth of Clyde in January 1736, the eldest of eight children, five of whom died in infancy. His father, also James Watt, seems to have had a distinctly varied career, being described at various times as chandler, carpenter and joiner, shipwright and ship owner. He was clearly an important person in the community, being appointed as baillie, or chief magistrate, in 1751. His wife, Agnes, was a formidable woman and intellectual who had family connections with Glasgow University that were to be useful to her son in later life.

As a boy, James suffered from a good deal of bad health, often involving severe headaches, a problem that was to plague him for much of his life. He was taught at home for a time by his mother before going to a primary school and then on to Greenock Grammar School. The education at that time concentrated very much on the classics, and young James was taught Latin and Greek, but apparently showed little enthusiasm for either subject. The teachers assumed that he was not a very bright pupil, but for a boy who was already showing an interest in all things mechanical, the classics

Image: James Watt. (Jennifer Tann)

probably seemed of little use. The one subject that held his attention, and in which he proved far from stupid, was mathematics. In his spare time, he liked to help his father in the workshop and made models. His formal education came to an abrupt end when his father's business suffered a severe financial loss. It was time for the boy to start earning a living. It was decided that, with his interest in mathematics and his practical skills, his ideal occupation would be as a maker of scientific instruments.

In June 1754, he arrived in Glasgow, where he stayed with members of his mother's family, the Muirheads. George Muirhead was chair of the Humanities department at Glasgow University. He was able to introduce Watt to a number of members of the Glasgow Literary Society, including Joseph Black, a distinguished scientist who made major contributions to both physics and chemistry. His researches into heat were to prove important later on in Watt's career. Another academic he met at that time was the Professor of Natural Philosophy – the eighteenth-century term for what we would now simply call science – Dr Robert Dick. They were to be lifelong friends, and it was Dick who persuaded Watt that, if he wanted to be a really successful instrument maker, he needed a better teacher than any available in Glasgow. He suggested that he move to London, and gave him an introduction to James Short, who was famous for his very high-quality reflecting telescopes that were bought for observatories all round the world.

Watt arrived in London in June 1755 and managed for a while as an assistant to a watchmaker called Neale. Then Short introduced him to a scientific instrument maker, John Morgan, with a workshop and office in Cornhill. He agreed to take the young man on and teach him for a fee of 20 guineas for a year, during which time he would work as an unpaid assistant. It was a real opportunity to learn from an excellent technician, but it came at the price of considerable hardship. Watt was keen to cost his father as little as possible, but with no income he had to rely on money sent from Scotland and live as frugally as possible. His health had never been good, and hard work, long hours, poor accommodation and little food did nothing to improve it. However, he was able to declare that, after a year, 'he was able to work as well as most journeymen'. Although he had the highest regard for Morgan, he was simply not able to continue living in such conditions and, after purchasing a few tools, he returned to the family home and months of convalescence.

Health regained, he returned to Glasgow, but had trouble getting work. He may have considered himself as good as any journeyman, but he had not completed an apprenticeship, so was not qualified. And as his father was not a burgess, he had no specific rights to help establish a business. He was lodged at the university, where he resumed his friendship with Robert Dick and helped him organise a collection of astronomical instruments that had arrived from Jamaica. His influential friends managed to get him a workshop, and he became instrument maker to the university. The distinguished physicist John Robison recalled meeting Watt in 1758:

I saw a Workman and expected no more – but was surprised to find a philosopher as young as myself, and always ready to instruct me. I had the vanity to think myself a pretty good proficient in my favourite Study and was rather mortified at finding Mr Watt so much my superior.

He was not the only academic to be impressed by Watt's natural abilities. Joseph Black employed him in making instruments for his scientific studies. Watt also built an organ and he and Black produced a perspective machine, a device to enable artists to create true perspective in their drawings. It consisted of a pentagraph and an easel, and several were made over the years. Like Robison, he found Watt to be far more than a mere workman, describing him as 'a young man possessing most uncommon talents for mechanical knowledge and practice'. Black was the first to formulate the idea of latent heat, a theory that describes the heat that has to be exchanged when a body changes from one state to another, for example when water turns to steam at constant temperature and pressure. It was a theoretical concept that was to have a very definite practical application. What is not clear is whether or not Watt ever discussed it in detail with Black and understood its implications.

After working for a year from the university, Watt decided that he would specialise in making Hadley's quadrants, navigational instruments used to measure the altitude of the sun to establish a ship's position of latitude. He thought he would need to travel to busy ports, especially London and Liverpool, to sell them. He did, however, find ready sales in Glasgow as well. In order to capitalise his new business, he went into partnership with John Craig, a gentleman about whom we know very little, but is believed to have been an architect. By 1763, the business had prospered to an extent that he had employees and apprentices and had moved to Trongate in the centre of Glasgow. As well as now being able to offer a wide range of instruments and not just quadrants, he sold 'toys'. These were not playthings, but small articles, such as metal buckles and buttons, many of which came from the most important toy manufacturer in Birmingham, Matthew Boulton – a connection that was to prove vital in his later career. In 1764 Watt married his cousin, Margaret Miller, and they had two children, Margaret, born in 1767, and James, born in 1769.

Watt first became interested in steam engines when John Robison suggested they should investigate the idea of using steam power to drive a road vehicle. This is somewhat ironic, given Watt's implacable opposition to just such a development later in his career. In any case, the idea was never pursued. He was also interested in developing a device based on Papin's digester, an early form of pressure cooker using steam at high pressure, but he was worried that the pressure would be too great and was likely to cause severe damage if the vessel burst. He was to retain his aversion to the use of high-pressure steam throughout his career.

He might have lost interest in all things steam if the university had not sent him a model of a Newcomen engine that did not work as it should. As Smeaton and others had discovered, Newcomen engines were notoriously inefficient; in fact in terms of thermal efficiency they worked generally at a miserable 1 per cent – in other words only one hundredth of the heat supplied was converted into useful work. But so far no one had done more than make improvements without questioning the basic principles.

Watt calculated what the engine's performance should have been and then tested it to find out how that figure compared with its actual performance. He discovered that more steam was being used than he expected and more water was also being used for condensation. He realised that the extra energy was being used in changing the state of matter – water into steam and steam back into water – in other words, due to latent heat. He described the moment when the solution to the problem came to him:

It was in the Green of Glasgow. I had gone to take a walk on a fine Sabbath afternoon, early in 1765. I had entered the Green by the gate at the front of Charlotte Street and had passed the old washing-house. I was thinking upon the engine at the time, and had gone as far as the herd's house, when the idea came into my mind that as steam was an elastic body it would rush into a vacuum, and if a communication was made between the cylinder and an exhausted vessel it would rush into it, and might there be condensed without cooling the cylinder.

In other words, his idea was not to condense the steam in the cylinder itself, but in a separate condenser. It was an idea that would revolutionise the development of the steam engine, but it was not yet a complete solution. One extra device would be needed, an air pump to draw off the condensate from the 'exhausted vessel'. He realised that the pump could be worked mechanically as part of the engine itself.

Watt began thinking of ways in which he could apply his new idea to create an improved steam engine. 'In the course of one or two days the invention was thus far completed in my mind, and I immediately set about an experiment to verify it practically.' At this stage, he was still thinking in terms of an atmospheric engine with an open-topped cylinder in which air pressure would drive the piston down. His first experiments were all with models, but the time had come to move to actual machine construction.

Professor Black introduced him to Dr John Roebuck, who had originally trained as a doctor of medicine and had practised in Birmingham, where he became friends with Matthew Boulton and others in his circle. These included Dr William Small, another medical practitioner, but one who took a wider interest in all the sciences. This encouraged Roebuck to begin experimenting himself and then when these were successful he started to look for ways to develop his results for profitable uses in industry. One of his successes was a new process for making sulphuric acid, and he began manufacturing it at Prestonpans in Scotland in partnership with Samuel Garbett. For a time the new business prospered but, as he had failed to take out a patent for his process, the company began to suffer from competition.

Roebuck now turned his attention to a new industry. In 1759, he again went into partnership with Garbett and William Cadell to build an iron works on the River Carron near Falkirk that would use coke firing for the furnaces, as established at Coalbrookdale by the Darby family. Technically the works became very efficient, but suffered badly from under capitalisation and were soon running up massive debts; ultimately Roebuck sold his share in the company. Roebuck had now moved to Scotland and begun a new venture developing coal mines at Kinneil near

Bo'ness on the Firth of Forth. The mines gave trouble from the start. First the coal was not of the quality he had hoped, and the old Newcomen engine proved inadequate in preventing frequent floodings. He sought help with his growing problems by appealing to his old friends in Birmingham, including Boulton, but none were very interested in getting involved with Scottish mines. It was now that Roebuck called in Watt to see if one of his improved atmospheric engines would be more successful in draining the mines. He was impressed by Watt and his idea, and saw another money-making opportunity.

Black had helped finance Watt's experiments with steam engines, but now Roebuck offered to repay all the debts and take over financing Watt and help him obtain a patent. In return he wanted two thirds of the profits. Watt was attracted to the idea, and suggested that the best way forward would be to have him settled in a small house at Kinneil, where he could work on the specifications for the patent and develop models. But steam engines were not his only concern. He had been called in to help with the survey of the Forth & Clyde Canal and in 1767 he had to go to London to give evidence on the Parliamentary Bill. On his way back to Scotland, he visited Dr Small and Erasmus Darwin and, although Boulton was not at home, he was given a tour of his factory, the Soho Works. The steam engine was discussed, and Small wrote to Watt shortly afterwards to say that he had not known of the Roebuck connection, but had discussed the whole steam engine project with Boulton and had suggested a partnership, provided Watt was willing to move to Birmingham.

Meanwhile, Watt was working on a new idea, a steam wheel. This was a circular tube with steam inlets and outlets, and the steam was used to push against freely moving weights inside the wheel. As the weights moved, they would cause an imbalance and the wheel would spin. He spent a great deal of time on the project, which turned out to be totally impractical. Watt made another trip to London, but in August 1768 he was back at Birmingham and this time was able to spend time with Boulton; there was an immediate rapport between the two men. Back in Scotland, Watt explained that Boulton was still interested in the steam engine. Roebuck was once again decidedly short of funds, so he wrote to Birmingham offering a share in the invention and the profits from the sales in three counties around Birmingham. Boulton rejected this limited idea and, in one of the classic statements in industrial history, he wrote to Watt to explain what he would hope to do: 'My idea was to settle a Manufactory near to my own by the side of our canal where I would erect all the conveniences necessary for the completion of the engines and from which Manufactory, we would serve all the world.'

Watt was still in Scotland and tied to Roebuck, but with occasional engineering projects including surveying the line of what would later become the Caledonian Canal. It was while he was involved in this project that he received news that his pregnant wife was dangerously ill. He hurried home, but the child was stillborn and his wife had died on 24 September 1773, two days before he arrived. The 1770s were a bad time for many, with a trade depression under way, and among those who suffered was Roebuck, who owed money to many concerns, including Boulton. He offered to sell a half share in his rights for the steam engine, but Boulton was not interested. In order to make a real difference he needed to be the sole partner with Watt. But as Roebuck's financial position deteriorated, he eventually reached the point where he had no option: he agreed to sell all his rights to Boulton. The partnership of Boulton and Watt was finally to become a reality in 1774.

Watt moved to Birmingham and was now able to devote all his energies to developing the steam engine. The separate condenser was the key element, but at first it was still only a way of making the old atmospheric engine more efficient. Watt realised it was less than perfect. One problem still remained: the loss of heat through the open-topped cylinder. But, if he closed the top, then air pressure could not be used to move the piston. Atmospheric pressure at sea level is roughly 15psi and Watt realised that it was perfectly possible to produce steam at that pressure or even higher. He would not need air pressure at all. He could close the top of the cylinder and use steam pressure to move the piston. He could also reduce heat loss by encasing the cylinder in a wooden jacket. The atmospheric engine had become a steam engine and, unlike the Newcomen engine, was capable of development in all kinds of ways. He had already taken out a patent in 1769, but this was about to expire, so in 1775 an application was made for an extension to 1800, but this time the patent was so loosely worded as to effectively prevent anyone developing steam power that involved a separate condenser, an air pump and using the expansive power of steam. The new steam engine was to become the driving force of industry and transport throughout the nineteenth century and beyond.

All the precision parts for steam engines were provided by Boulton & Watt, while accurately bored cylinders were supplied by John Wilkinson from his ironworks at Bersham in North Wales. Other parts had to be supplied by the purchasers, but were carefully vetted by Boulton & Watt. The engine parts came with a complete set of instructions on how everything was to be assembled. The first new engine was installed at Bloomfield Colliery at Tipton in the Black Country and started work in March 1776. The following year saw two important events for Watt. He remarried and the first engine was sent down to Cornwall to be installed at the Wheal Busy mine. Cornwall was to prove vital to the whole success of the business over the next decades. The profit came from a premium that was paid of one third of the saving made in changing from an atmospheric to a Watt steam engine. The Newcomen engine improved by Smeaton had a duty rating of 7 million, but the earliest Boulton & Watt was rated at 22 million, though that figure would later be improved. Between 1777 and the end of the century, some fifty engines were sent down to Cornwall, with cylinder size ranging from a modest 18in to a substantial 64in.

Cornwall was so important to the company that it decided it need a man on the spot to look after its interests. The man selected for the job was a Scotsman, William Murdoch. The story of how he got the job may not be true but it is certainly different. He walked from his home in Scotland to Soho, where he called on Boulton, wearing a wooden hat. This was so odd that Boulton wanted to know more about it and Murdoch explained he had made it himself to demonstrate his skill. Boulton employed him.

Both Boulton and Watt made regular trips to Cornwall. Watt took his new wife, Ann, with him, but she was not impressed. She enjoyed the coastal scenery but hated the mining areas and the Cornish – 'a set of laziest wretches that breathes on the earth'. By 1780, however, they were at least able to rent a comfortable house near Truro. Watt also had problems during their visits as he no longer had a secretary to copy out his correspondence. He decided to avoid the tedious task of copying his own letters by devising a copying machine. He began by thickening the ink with gum Arabic, so that when a sheet of thin tissue paper was pressed against it, an impression was formed. It was, of course, reversed, but by holding up the back of the sheet to a light it could be read. The next step was to mechanise the process by feeding the papers through rollers. Boulton saw the commercial possibilities and a new company was formed,

James Watt & Co., and a patent obtained in 1780. It was a success, with 630 of the copying presses being sold at 7 guineas each. Its usefulness, however, came to an end early in the nineteenth century with the invention of carbon paper.

Watt now had a new problem to solve. Boulton was very keen on the idea of a steam engine that could produce rotary motion; the steam wheel had been abandoned, so it was a case of adapting the existing engine. The snag was that, with the existing engine, the piston was attached to the overhead beam by a chain and, although steam could be used to raise the piston, the chain would be useless for pushing upwards. It was not possible simply to replace the chain with a metal rod, because the end of the beam

6.1 A Boulton & Watt beam engine: Watt's parallel motion can be seen on the left, attaching the piston rod to the beam. (From Smiles, *Lives of the Engineers*, 1862)

moved through the arc of a circle. Watt's answer was the parallel motion. A rod from the piston was fastened to one corner of a shifting parallelogram of rods attached to the beam. The connecting rod was thus kept more or less to a vertical motion. Watt always regarded this as his greatest invention and it is a most satisfying device to see in motion.

This solved one part of the problem, but there still remained the question of how to convert up and down motion into circular motion. There was an obvious answer: the crank. But there was an existing patent covering that, taken out by James Pickard, so Watt needed an alternative. Several ideas were tried, but the one he eventually settled on was known as the sun and planet gear. A sweep arm was attached to the end of the overhead beam, on the opposite end of which was the 'planet', a small cog that ran round and engaged with the 'sun', a larger cog mounted on a shaft. Just as with the older water wheel, once you have a turning shaft it can be used to power all kinds of machines, either through cogs or belts. The sun and planet only remained in use until the crank patent expired.

There was to be one further development of importance. In 1788, Boulton had visited the water-powered Albion flour mills, where he had seen how the movement was kept regular by means of a centrifugal governor. As the flow of water to the mill was irregular, changes in the flow affected the smooth running of the machinery. The governor consisted of a pair of heavy balls connected to moveable arms: they turned with the water wheel, and the faster they turned, the further apart the spinning balls moved and that, in turn, worked on a spring, keeping the motion uniform. He wrote to Watt to tell him about the device, and the latter realised that it could be adapted to regulate the inlet valve of the steam engine to ensure smooth running. Watt now had a complete and versatile machine that could be used in many different ways. One area where the rotative engine was welcomed was in the rapidly developing cotton industry. In just a few years there were sixty-one engines at work in Lancashire and Cheshire and a further nineteen in the Midlands.

Everything was going well with the engine works and, at first, that was especially true of their Cornish connection. But Cornish engineers had a long tradition of going their own way and, although at first they had welcomed the improved engines, which were saving them a great deal of money, some of them felt they could make improvements. They had already done so with the Newcomen engine, and saw no reason why

6.2 Watt's sun and planet gear, used to transfer the up and down motion of the piston into rotary motion. (Newtown graffiti)

they should not do the same for a Boulton and Watt. They did not regard the patent as any sort of obstacle, but Watt was determined not to let in any competitor. The first serious threat came from Jonathan Hornblower of Penryn, who had developed the first compound engine. Instead of exhaust steam from the cylinder being directed straight into the condenser, Hornblower recognised that it was still under pressure and fed it into a second cylinder, before finally being exhausted. As far as Watt was concerned, it was just his engine with a second cylinder stuck on the side and clearly in breach of his patent. In fact, the Hornblower engine never proved a great success – although the principal was sound – but that did not stop Watt from taking out an injunction against him. But then a more serious competitor appeared on the scene: Edward Bull.

Bull had actually been one of the engineers charged with overseeing the erection of Boulton & Watt engines, but it seems that he was not only unwilling to follow instructions, but was thinking of improvements of his own. This was too much for Watt, who informed the company's agent in Cornwall, Thomas Wilson, to have no more to do with him. But that was not the end of the Bull story, for he recruited a young mining engineer to join him in developing and installing a new type of engine. The newcomer was Richard Trevithick, whose father had been one of those who had tried to improve the Newcomen engine. Bull's engine certainly looked different. There was no overhead beam, but instead the cylinder was inverted and set directly above the shaft. The piston rod was then attached directly to the pit work controlling the pump. There was, however, no denying that the closed cylinder and separate condenser that Bull used were both covered by the Watt patent.

At first Watt was inclined to ignore Bull, on the basis that he was bound to fail. But when engines were built, they proved satisfactory and, according to Bull, performed better than his own, so Watt took action and sued. The court found against Bull, but there was a sting in the tail. Lord Chief Justice Eyre, in his summary, agreed there was a clear case of piracy, but cast doubt on the validity of the Watt patent: 'Letters patent can be of no avail unless there is a true Specification of that invention. I confess I have myself very grave doubt whether this Specification is sufficient.' That was to prove all too accurate a statement later when the patent was reviewed with a view to extending it.

There was now an injunction against Bull installing his engines, but no mention was made of Trevithick. He decided to carry on and install a Bull engine at Ding Dong mine, on a high ridge overlooking St Michael's Mount. Watt was unable to prove that Bull and Trevithick were actually partners, so to stop the work going forward they needed to serve a new writ on Trevithick. That proved to be far from simple. The bailiff who arrived at Ding Dong to serve the papers was grabbed by a group of miners, who tied a rope around his waist and suspended him over the shaft, enquiring politely if he really wanted to serve the papers. He agreed that, on consideration, he would rather not. But Trevithick became too bold and actually visited Birmingham, where he was spotted and the papers were served. The legal battle had been lost. In fact, the Bull engine never really lived up to expectations. One result of the legal case of 1793 was the doubt cast on the patent, and in 1800 it was not renewed.

6.3 The Boulton & Watt factory beside the Birmingham Canal. (Jennifer Tann)

In 1794, a new company was formed, Boulton, Watt & Sons, the latter being James Watt Jnr and Matthew Boulton Jnr. It was decided that this was the ideal time to provide what Boulton had wanted from the beginning, a new factory for the manufacture of engines. The site chosen was beside the Birmingham Canal on what is now known as the Soho loop. Work began on building the foundry in 1795 and it was opened in January the following year.

Before the patent had been lost, another rival appeared, this time one of the company's most trusted employees, William Murdoch. He had been experimenting with the idea of a steam-powered vehicle, and had even built a working model that he had tried out at night after work. The sight of the machine with its smoke and flames terrified the local pastor, who was convinced he had seen the devil himself. By 1795, he was ready to apply for a patent when Watt and Boulton heard of his plans. They made it very clear that, if he went ahead, he would no longer be working for the company, so he abandoned the project and settled for a reliable income

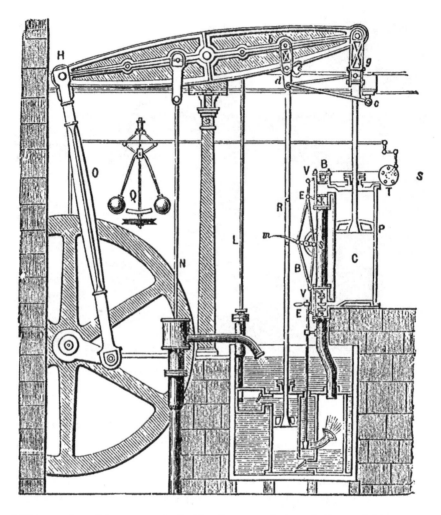

6.4 Later versions of the steam engine had iron beams, cranks instead of sun and planet gear and centrifugal governors, marked 'Q' in the diagram.

instead. He left Cornwall for Soho in 1798, where he began working on a new idea – gas lighting. In 1802 it was used to illuminate the factory.

Fighting piracy was important to Watt, but it only took up a small part of his time. He was constantly looking for improvements in the engine, and he had a busy and interesting social life. He was a member of the Lunar Society that met regularly at Boulton's house. It got its name because, in the absence of street lighting, meetings were held when there was a full moon to light the way. Many leading scientists and industrialists were among the members, including Josiah Wedgwood, Erasmus Darwin and Joseph Priestley. It helped Watt to develop many of his own scientific ideas, not necessarily connected with the engine business. For example, he patented a smoke-consuming furnace in 1785 to help reduce air pollution in industrial towns – a problem that he must have been aware was being made worse thanks to his own machines.

Watt retired in 1798 and bought a farmhouse in Doldowlod in Radnorshire, which was converted into a fine country home, where he often spent the summer. He and his wife enjoyed travelling and, even in retirement, he continued working on inventions, not with any aim of making money, but simply for the pleasure of doing so. He did take on one commission. He was asked by the Glasgow Waterworks if he could help solve a problem they had encountered: how to get filtered water across the Clyde to a pumping station at Dalmarnock. He sent the company detailed drawings of a flexible water pipe, consisting of several joined sections. He refused payment, but the company were so pleased with the result that they presented him with a silver dinner service. In his later years, he was working on a machine that could be used to faithfully copy three-dimensional objects, such as sculpted busts. He was still working on the idea when he died at Birmingham on 25 August 1819. Very appropriately, he was buried next to his friend and partner Matthew Boulton at St Mary's, Handsworth.

The company the two men had founded continued to prosper after their deaths, with James Watt Jnr in charge. They did exactly what Boulton had said he wanted to do: they continued to supply engines to the whole world. Watt's own contribution is now universally remembered in the scientific standard unit of power that bears his name.

7

RICHARD TREVITHICK

We have already met Trevithick in his role as a steam pirate, but we have to backtrack a little first to look at his earlier life. His father, also Richard, was a highly respected mine captain; that is to say, he was responsible for overseeing all the work at a mine. He was also agent for the powerful landowning family, the Bassets of Tehidy. William Basset would later become Lord de Dunstanville. The son was born in a house in the heart of the mining area, between Pool and Carn Brea on 13 April 1771. Shortly afterwards, the family moved to Penponds, near Camborne, a house that still survives and can be seen to be both comfortable and having a touch of elegance in its panelled rooms. He had four elder sisters, who might have spoiled him if another daughter, Thomasina, had not arrived later. His education was rudimentary and involved learning by rote. He was unimpressed by the teaching. When he was told off for doing sums by the wrong method, even though he got the right answer, he simply told the teacher that he could do six sums to his one. He was far more interested in the practical work of the mines than

Image: Richard Trevithick. (Francis Trevithick biography)

7.1 The thatched house where Trevithick was born, surrounded by the workings of South Crofty mine. (Royal Institute of Cornwall)

in book learning and by the time he was 15 years old he was working at Dolcoath mine, one of the most important in Cornwall. He was paid what was then a handsome sum for a boy of £1 4s per month, which was soon increased to £1 6s.

He rose rapidly in his chosen profession and, as we have seen from the last chapter, he was soon teamed up with Bull in devising a new type of engine. In a way he was merely following in his father's footsteps, for he had also infuriated James Watt, who wrote of the 'impudence, ignorance and overbearing manner of the man'. After losing the battle to build engines, Trevithick returned to working for mining companies. His work regularly took him to the foundry run by the Harvey family at Hayle, which was kept busy providing the iron work for steam engines. His contact was William West, who was married to Joanna, one of the Harvey daughters. They became good friends, and Trevithick was introduced to the younger sister, Jane. The Harveys were a bit too grand to welcome the idea of the reckless steam pirate becoming a member of the family, but in August 1797 Trevithick's father died, and the son took over his responsibilities. He was now a recognised first-class engineer and enjoyed a considerably increased income. He and Jane were married in November the same year and moved to a house in Redruth.

7.2 The ruins of Ding Dong mine, where Trevithick and Bull experimented with their inverted engine. (Simon Cook)

Trevithick's improved position did not prevent him from working on new ideas. His first invention was not concerned with the steam engine itself, but with the pumps that it worked. In the pumps generally in use, the steam power was needed to lift the water as the bucket pump rods were lifted. His idea was the plunger pole, which forced water up through an H-piece into the pump column. The water was raised in a series of lifts, generally of 30 fathoms (180ft) each. The working stroke was powered by gravity, and the engine was only required to lift the rods back up again. It was a success, and many were installed, though Trevithick only made money from those he worked on himself. He now began to think about ways in which the process could be reversed: if a plunger pole could lift a column of water, then a column of water could be used to move a piston. The idea was not new, but had never been seen in Cornwall, so it is unlikely Trevithick had ever heard about it. He installed a simple version at Roskear in which water under pressure was used to raise a pole in the cylinder, which then dropped back under its own weight when the water flow stopped. Unlike the pump, it was not widely adopted in Cornwall, where water supply was a problem, but was taken up in other parts of the world, notably in the lead mining area of Derbyshire. As was often to prove the case in Trevithick's life, he was much better at inventing than he was at business, and he often failed to get the money he felt was due to him.

By the end of the 1790s, Trevithick began thinking about steam engines again. While the Watt patent lasted, every engine had to conform to his ideas, which meant using steam at low pressure. If more power was needed then you ordered a bigger engine: the largest supplied to Cornwall by Boulton & Watt had a massive 8ft diameter cylinder. This was fine for the slow, pumping engines, but the mines were also installing whim engines to haul material up and down the shaft. Trevithick wondered if, by using high-pressure steam, he could build, smaller, faster engines. But he was uncertain about how this would work if he had to use a separate condenser. He consulted a scientific friend, Davies Gilbert, on what would be the effect of working without the condenser, and how much power would be lost. He was delighted to learn that it would only be a maximum of one atmosphere, and he would also be able to do away with the air pump.

He began work with a model, using his kitchen table as a laboratory bench. The model was built by his brother-in-law, Samuel West and, when the time came to try it out, Gilbert Davies was given the job of stoking the fire below the tiny boiler and Lady de Dunstanville was given the honour of turning open the valve to set it in motion. It was a success. The high-pressure engines went into production and became known as 'puffers', as the steam escaped into the atmosphere with a roar and a whistle. The first engine went to Ding Dong, where Bull had erected his engine, and soon there were fourteen engines erected in Cornwall. Boulton & Watt were alarmed, but with the end of the patent in sight, and with no guarantee it would be renewed, they decided against going to court and instead proposed a trial between one of their engines and a Trevithick puffer. Trevithick was so confident that he placed a bet of £50 on his machine – and he won.

In order to supply high-pressure steam, he needed something better than the haystack boilers then in general use, which were little more than oversized kettles. His new design was circular, with an internal flue. A firebox was set at one end, from which the hot gases passed through a U-bend to arrive at a chimney next to the firebox. A comparatively small cylinder was used, and the exhaust steam was released through a blast pipe that inevitably became very hot. However, Trevithick used it efficiently by surrounding it with a water pipe, which heated the water and could then be returned as boiler feed water. His next step was to make light versions of the puffers, which he mounted on wheels, so that they could be pulled by horses to wherever they were needed.

So far, Trevithick had produced two useful inventions that improved the efficiency of machines already in use. He was now about to make an altogether more radical change. He had set his puffers on wheels and now he wondered whether if he used the power to turn those wheels, the whole thing would be able to move itself along. It depended on whether there was sufficient friction between the wheels and the ground, so he experimented with an ordinary cart, moving the wheels by hand – and the cart moved. He was now ready to develop what would become a steam locomotive. He was not the first to have this idea. As early as 1769, a former officer of the Austro-Hungarian Army had designed a steam tractor that he intended for use in moving artillery pieces on the battlefield. It proved to be a very clumsy device that was difficult to control and, because it had no feed pump and there was little space for water, it could only run for twenty minutes. After that time it would be stopped, allowed to cool down, and then be fired up again. No one was interested and the project was abandoned, and it is doubtful if news of it ever reached Cornwall. As mentioned earlier, Murdoch had built a model steam locomotive, but he too had given up any further development when faced by Watt's implacable opposition.

Now it was Trevithick's turn, and he too began by building a model before arranging for a full-scale engine to be built. The actual construction involved several people. The cylinder and boiler were cast at Harvey's, while complex and detailed parts were made by Trevithick's cousin, Andrew Vivian, and assembled by a Camborne blacksmith, Jonathan Tyack. There are no working drawings, but Trevithick's son, Francis, who wrote his biography, produced what he considered the most likely arrangement. There was a basic return flue boiler as in the puffers, with a single cylinder set into the top. The piston rod was attached to a crosshead, from each end of which connecting rods went down to cranks on the rear wheels. There is no indication of how it was steered, although a contemporary spoke of a 'steering handle'. A replica was, however, built to celebrate the bicentenary of the first trial and this used a tiller to turn one pair of wheels, which worked quite well but was far from perfect.

On Christmas Eve 1801, the engine was steamed and, according to an eyewitness account, as soon as it started moving, seven or eight jumped

7.3 Trevithick's original model for the first steam carriage. (Steam Museum, Straffan)

on for the ride. The engine stormed up Camborne Hill at faster than walking speed and, after travelling for nearly a mile, was turned around and steamed back to the middle of Camborne. The event is still remembered in the Cornish song: 'Going up Camborne Hill, going down'. On Boxing Day, Trevithick and Vivian decided to take the engine up to Tehidy House to show it off to Lord and Lady de Dunstanville. Before it got there, it hit a watercourse and tumbled over. As the accident happened near an inn, Trevithick and Vivian decided to abandon the engine for the time being and go and enjoy a drink. They forgot to extinguish the fire in the firebox, the boiler became red hot and exploded. That was the end of the first road locomotive.

Trevithick was not unduly worried. It had always been an experimental prototype, and now he had a much more ambitious plan. He would design a steam carriage that would carry passengers. By march 1802, Trevithick and Andrew Vivian had a patent for a steam carriage. It was very different from the Camborne engine. The steam cylinder was now horizontal and the piston rod forked to allow for the movement of the crank shaft that drove the rear wheels. The wheels themselves were enormous, roughly 8ft in diameter, which meant that the carriage itself was high above the ground. Steering had been a problem with the Camborne engine, and the method used for the carriage was not a great improvement. Once again, a tiller was used, this time to manipulate the single, small front wheel. The iron work for the engine was cast at the

Harvey works in Hayle, but the actual carriage work went to William Felton, at Leather Lane in Clerkenwell, London. It was a capacious affair able to hold eight passengers.

Demonstration runs were made in London, but amazingly, given this was a unique machine, seemed to attract very little public attention. Everything did not always go according to plan. One account describes a journey on the carriage with Trevithick himself in charge. They went along Tottenham Court Road and round Euston Square, when the carriage went out of control and tore down several yards of railing from a garden wall, at which point an angry owner opened a window and shouted: 'What the devil are you doing there! What the devil is that thing!' There was one report of a triumphant run along Oxford Street when the shops were closed, and all the upper windows crowded with cheering spectators. What was it like to travel in the steam carriage? Another member of the Vivian family, a former sea captain, declared that, after one jaunt, he was more likely to be shipwrecked on the carriage than he had ever been on his ship, and refused to go again.

7.4 The replica of Trevithick's London carriage, built by Tom Brogden. (Tom Brogden)

As with the Camborne engine, a replica was made, this time by Tom Brogden, and the carriage was brought down to London to Leather Lane for the unveiling of a plaque to mark the bicentenary of the construction. A descendant of the engineer, Frank Trevithick Okuno, unveiled the plaque and I was invited to say a few words at the ceremony, after which Frank and I were treated to a drive around Regent's Park. Based on that experience I can report that it was very comfortable and there were no steering disasters.

Trevithick's hopes of raising money for a whole fleet of steam carriages were soon dashed. There was no enthusiasm among investors and, although it worked well enough, not all the problems, notably steering, had been solved. There was alarm that a vehicle with a fire underneath might explode, and an argument that would appear again with all forms of steam engines emerged for the first time: it would frighten the horses. Reluctantly, Trevithick was forced to abandon steam road carriage development forever. But he had not given up entirely on the idea of moving vehicles by steam.

There is a certain amount of doubt about what happened next in this story. We do know that in August 1802 Trevithick was at the famous Darby iron works in Coalbrookdale, making a series of tests on an engine with ever increasing steam pressure. Considering that with the Watt engines pressure was rarely allowed to rise above 10psi, here we find Trevithick working with pressures as high as 145psi. This in itself is remarkable, but what is really interesting is a postscript to a letter he wrote describing his experiments, in which he notes that they had ordered a steam locomotive to work on their railway. Nothing is known about this engine, but there is confirmation that it existed. W.A. Reynolds, nephew of the iron master William Reynolds, remembers being given a wooden model of the engine, which he took to pieces – 'an act which I now repent as if it had been a sin'. A visitor to Coalbrookdale in 1884 was shown the cylinder of Trevithick's engine, which was 'cherished as a valuable relic'. It is now thought that a drawing thought to represent a later engine was indeed of this locomotive, which would be the very first locomotive to run on rails. The details of this engine, a 4¾in cylinder with a 3ft stroke, fit well with the drawing. We have no idea how well it performed, but his next engine attracted far more attention.

In 1803, Trevithick received a letter from Samuel Homfray of the Penydarren iron works at Merthyr Tydfil inviting him to visit to discuss the possibility of using a high-pressure steam engine to solve a transport problem. At the time four companies dominated the highly successful South Wales iron industry: Penydarren, Dowlais, Plymouth and Cyfarthfa.

Transport for the various works was largely dependent on canals, and the Glamorganshire Canal served both Penydarren and Richard Crawshay's Cyfarthfa works. The two companies were in constant disagreements about which should have right of way at the locks, and this was particularly true of the heavily locked upper section between Merthyr and Abercynon. Homfray decided to get round the problem by bypassing that section with a tramway, constructed out of 3ft-long metal rails set 4ft 4in apart and built with a gentle gradient of 1 in 145. Soon after meeting Trevithick, Homfray was ordering an engine to work on the tramway instead of horses. It was to be a versatile engine, for when it was not in use on the tramway it could be used as a stationary engine to work machinery at the foundry.

Crawshay, who seems to have disagreed with Homfray on practically everything, at once declared that the whole idea was preposterous and iron wheels would never be able to get a grip on iron rails, and backed up his assertion with a £500 wager. For Homfray to win he had to provide a locomotive that would take 10 tons of iron from Merthyr to Abercynon and return with the empty trucks. This was a huge bet, equivalent to over £10,000 at today's prices, and gives an indication of the wealth of the iron industry at that time.

The engine was constructed at Penydarren and was probably very similar to the Coalbrookdale engine. There was a wrought iron return flue in a cast iron boiler, with the firebox next to the chimney. An 8¼in diameter cylinder was set horizontally into the boiler and the piston rod connected to a crosshead reaching across the front of the boiler. From there the drive was taken to gears to the two wheels on the left-hand side of the boiler. As it was also intended for use as a stationary engine, it was fitted with a large flywheel. The boiler feed pump seems to have been intended to supply hot water to the boiler, but quite how this was achieved is uncertain. There is no indication of any braking system. This might seem strange, even foolish, but the replica Coalbrookdale engine now running at the Blists Hill open air museum does not have one either. I had the privilege of riding on the footplate and discovered the system used there. Steam was temporarily shut off and then, as the engine slowed, the gear was put into reverse and steam readmitted. It worked perfectly well.

There were several trials of the engine, in one of which it successfully hauled 10 tons of iron in wagons, and with a reported sixty or seventy people riding on them as well, at a stately 4mph. In February 1804 the day was set to test whether or not the bet had been won.

Distinguished observers agreed the engine had done exactly what was required, but Crawshay still refused to pay, raising some technical quibbles. In the meantime, the engine was also being put through its paces at the works, where it was used to work a hammer that would normally have been powered by a water wheel. It seemed a resounding success, but a major problem arose, not with the engine itself, but with the track. The engine proved too heavy for it and many of the brittle cast iron rails were cracked. The problem was never solved, but the engine continued in

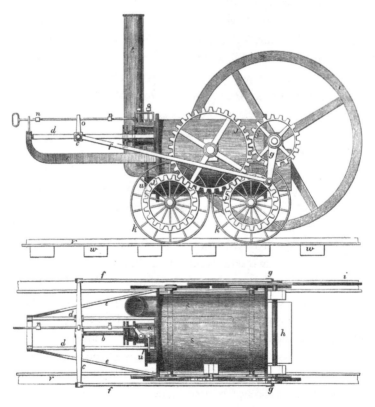

7.5 Trevithick's drawing for an engine for Gateshead, believed to be identical to the Penydarren engine. (From Francis Trevithick's biography of his father, 1872)

use as a stationary engine for many years. Trevithick received one order for another locomotive from Christopher Blackett of Wylam Colliery on Tyneside, but that suffered from the same problem; the rails, in this case wooden, could not take the load. It too was put to use as a stationary engine. Blackett, however, was later to request Trevithick to supply another engine after he had relaid the track with metal rails, but by this time Trevithick had made one last attempt to gain support with his final locomotive; when that failed to attract investment he abandoned locomotive work altogether.

That last attempt had taken him back to London, where he had first demonstrated his steam carriage. His new engine was named *Catch Me Who Can* and was very different from its predecessors. The cylinder was set vertically in the boiler and provided a direct drive to the rear wheels without the need for any gearing and the flywheel was no longer needed.

7.6 A ticket for visiting the London demonstration of Trevithick's last locomotive. (From Francis Trevithick's biography of his father, 1872)

Trevithick had a circular track built, surrounded by a high wooden fence, and the public were invited in to see the engine and, if they felt brave enough, go for a ride. The admission price was one shilling. The engineer John Isaac Hawkins was one who took up the offer and timed the engine as running at 13mph. He was told by Trevithick that it would be capable of 20mph on a straight track. But, once again, the track let him down, and the engine was taken off its wheels, sold and installed in an ornate barge for the use of the Lord Mayor of London.

Although locomotive business never transpired, the Trevithick–Vivian patent covered stationary engines as well, and this proved altogether more profitable. One area where the engines were used was in the foundry, working with wrought iron. There were two stages in the production. The first involved heating cast iron in a puddling furnace, one in which the fuel used for heating and the metal never came into direct contact. What emerged when the molten metal cooled was a spongy ball, full of small holes, in which liquid slag had been trapped. The slag was beaten out under heavy hammering to produce a 'puddled bar'. The bar was then reheated and passed through grooved rollers to provide either a square or round section bar. The new engines were used for both processes. In doing this work, Trevithick made some important discoveries, of which the most significant was a fuller understanding of the expansive nature of steam. In a letter of 1804, he described how he could improve the efficiency of the engine by cutting off steam to the cylinder before the piston had reached the end of its travel down the cylinder and allowing the existing pressure to complete the job. Variable cut-off for steam was to prove invaluable in locomotive development, but that still lay decades in the future.

It should all have gone well and made the partners a comfortable profit, but in the event it simply did not work out that way. As Trevithick of all people should have known, taking out a patent does not prevent others using it without payment or making adjustments to get around it. If he and Vivian had been sensible, they would have spent time consolidating the business, keeping check of performance, making amendments where necessary and generally keeping a step ahead of the competition. But that sort of routine work did not sit comfortably with Trevithick's personality: his mind was always looking for the next interesting idea, not worrying about what had happened to the last one. And Vivian turned out not to be the calm-headed businessman that was needed to maintain order and keep the inventive engineer concentrating on what would make a profit for them both. Trevithick had not found his own Boulton after all. Eventually the partnership was broken up and Trevithick was back on his own.

The ending of the Peace of Amiens in 1803, and the real possibility of a French invasion of Britain, enthused Trevithick with patriotic ideals. He proposed building a steam vehicle that would tow fireships into the heart of an invading French fleet. He was not only prepared to build a prototype but, if the method was to be used, he volunteered to be personally in charge. He was invited to discuss the idea with the Marquis of Stafford, who in turn introduced him to Lord Melville at the Admiralty. As was usual with Trevithick, once an idea was stuck in his brain he wanted to get straight on with it. He went up to Coalbrookdale, where a 10in steam cylinder happened to be available. This was put into a barge, and paddle wheels improvised out of two flywheels, with six planks arranged round the rim of each. The barge set off on a trial journey on the Severn, moving at a modest 7mph. The engineer, however, was to be disappointed. He was to find, as others had done, that great ideas could reach the Admiralty and then be sunk in a sea of bureaucracy, never to be seen again. Trevithick was soon, however, to face a new challenge.

The development of the new London docks and the cutting of the City Canal also brought a demand for other improvements to keep the waterways clear. Steam dredging was not entirely new: Boulton & Watt had built a crude steam-operated machine that worked what was in effect a giant ladle to scoop out mud. The first really successful dredger had been devised by an American, Oliver Evans. It had a continuous chain of buckets, worked by a small beam engine. Trevithick's dredger used the same basic idea, but with a high-pressure engine to work the chain. An illustration of 1893 shows what is clearly a small puffer on board a barge, with the chain of buckets set at one side of the vessel. It was successful, but was not really innovative. Now, having devised a machine to work on the Thames, Trevithick became interested in a proposal for burrowing under it.

In 1795, Parliament passed an Act allowing the newly formed Thames Archway Company to construct a tunnel beneath the river at Limehouse. The man in charge was to be another Cornish engineer, Robert Vaizie.

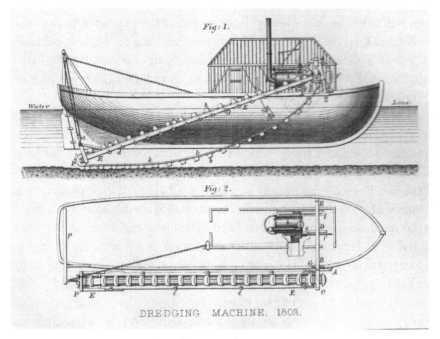

7.7 The steam dredger, worked by a 'puffer' mounted on the vessel. (From Francis Trevithick's biography of his father, 1872)

Initially, it would simply be a small drift to establish that a tunnel was possible and, when completed, it could act as a drain for the tunnel itself. Vaizie's plan was straightforward: sink a shaft on the bank to a depth well below the riverbed and then work outwards from the foot of the shaft towards the far bank. But things did not go well and, with all the first year's budget spent, all that was to be seen was a partly completed shaft and no work was even possible on the tunnel itself. Various eminent engineers were consulted about what had gone wrong and what to do next, but none of them could agree, so that did nothing to help. It was decided to call in a second engineer to work on an equal footing with Vaizie, and the man they chose was Trevithick.

Trevithick was, as usual, optimistic that all would go well. He was to receive £500 when the drift reached the halfway mark, and another £500 when it got to the far bank. This, he declared in a letter, would be money easily earned, with no risk to himself. At first his faith seemed justified. There had been an early ominous sign while sinking the shaft when an area of quicksand was discovered. But that work was completed and work began on the drift. It must have been very uncomfortable for the men, who had to work in a space just 5ft high and 3ft wide, but in the first week they drove 22ft. There were more worrying signs that the work would not prove straightforward as the men frequently encountered sand where they had expected more solid ground. Even Trevithick found this surprising and slightly worrying. In spite of the progress being made, there were delays and the proprietors, using a strange logic of their own, decided that as things had gone badly under Vaizie, but far better since Trevithick joined the team, any new delays must be due to the former. Vaizie was sacked and Trevithick given full control.

By December 1807, the drift had advanced nearly 400ft but then, instead of sand and gravel, they met rock. There was no question of blasting it away so near to the riverbed, so it had to be cut away with chisels and wedges. Worse was to follow as they got beyond that obstacle and started to find oyster shells mixed in the sand, suggesting they were much nearer the riverbed than they thought. Then they met quicksand and when, in January, they had just passed the 1,000ft mark, the roof began to break up and water flooded in. The hole was patched up and work continued.

At this point Trevithick was offered more money to get the work completed and persuaded his wife, Jane, to leave Cornwall and join him at the house he had taken at Limehouse. She must have been horrified to discover she was being brought to an area that could best be described as a slum. She had hardly had time to unpack before, on 26 January 1809, the tunnel roof totally collapsed and water flooded in. Trevithick managed to struggle out and arrived back, safe but without his boots and coated in river slime and mud. He at once began making plans for recovering the situation. At first, his thought was to sink a caisson down to the drift and use it to work inside to plug the hole. But then he had a much bolder idea. There was, he said, little point in continuing with the drift. It was time to dig the actual tunnel, but using a new method. A large caisson would be built and sunk down to the riverbed. Piles would be driven from inside it to the tunnel depth, and a 50ft-long section of tunnel created and lined with brick. The caisson would then be moved on and the next section completed. By now, however, the company had lost its initial enthusiasm: too much money had been spent with nothing to show for it. The whole project was abandoned.

Trevithick was not a man to dwell on failures for long. He promptly began thinking up new ideas. His first was for an improved paddle steamer that would be able to withstand the pitching and rolling it would encounter at sea. His idea was for a steam tug that would be used to bring sailing ships in and out of port. He put the idea to the Admiralty, which studiously ignored it – and as legislation did not allow any form of fire within dock areas, including fire to provide steam, nothing came of it. His next idea was for what he called a 'nautical labourer', which was a paddle steamer tug that would tow barges out to ships at anchor and, when alongside, would use a steam winch to unload cargo. In order to promote his idea, he took on a new partner, himself an inventor, Robert Dickinson. They took out a patent, but the new machine was boycotted by the workforce and the engineers threatened with drowning in the Thames if they ever tried to use it.

The partners now came up with a third idea. Liquid cargo for long voyages had been kept in barrels, but they proposed using iron tanks made to fit together snugly in the hold, saving a great deal of space. Once again, the Admiralty was approached and showed no interest, but Trevithick already had a new idea for his iron boxes. If you filled one of the boxes with water and threw it into the river it would sink. But if you pumped the water out, it would bob back up. The new idea was to attach several of these boxes to a sunken ship and use the method to raise the vessel. An experiment was made with a wreck off Margate. The operation was a success, but then the ship's owners said it was now up to Trevithick to have it towed back to harbour. He insisted on being paid for the towage: the owners refused and, with typical Trevithick bravado, he ordered the ropes holding the tanks to be cut and the ship sank back to the seabed. Trevithick got no pay – and the owners got no ship. Inevitably, the business started by the two men started losing money, the partnership was dissolved and Trevithick decided he had really had enough of London. It was time to return to Cornwall.

He was back working within a familiar environment and, at once, began working on ways to improve engine performance. He designed a new improved boiler for high-pressure steam to work big engines, and not just the small puffers. He used it in conjunction with a pumping engine that cut off steam very early in the stroke. An engine was installed at Wheal Prosper tin mine and later reports gave its duty as 26.7 million: the larger engines in use elsewhere could only manage a maximum of 22.3 million, and most were considerably less efficient. Trevithick was, as ever, quick to grasp new opportunities. He had visited Plymouth, where a breakwater was being built and where things were not going well. He at once set out to produce a new machine for preparing the stone, using a rotating bit instead of a chisel hammered home. And he was always on the lookout for new ways of using steam, including a threshing machine that worked far more efficiently than the horse-drawn machines then in use. He also designed what he called a plough, but was actually a rotary cultivator to tear up the soil rather than produce neat furrows. The latter was never built.

For once he seemed to be prospering when, in May 1813, he had a visit from Francisco Uvillé. Originally from Switzerland, Uvillé had come on behalf of a partnership working a silver mine at Cerro de Pasco, Peru, at an altitude of over 12,000ft in the Andes. They had hit a familiar problem – water flooding the works – and, to make matters worse, it seemed that the main deposits were to be found at even greater depths. Uvillé had been to see Boulton & Watt, but their massive engines would have been impossible to move up the rough mountain roads and he was unconvinced the low-pressure engines would do the job. So instead, he had come to Cornwall to see Trevithick's lighter, smaller but still powerful engines. He was so impressed, he ordered six engines to be sent to Peru, and Trevithick

7.8 The silver mines at Cerro de Pasco, c.1850. (Tony Morrison, South American Pictures)

arranged to have a team of men from Cornwall go with them to install them and set them to work. It seemed a fine commercial deal.

Troubles appeared after the engines reached Peru. The team had not been able to get them to work, so the only answer seemed to be for Trevithick to sail to South America in 1816, sort things out and come back home again. The first part of the plan worked well, but Trevithick had not paid any attention to South American politics. Simon Bolivar was in the process of advancing through the continent, liberating areas from colonial rule – and they arrived at the silver mines, which he promptly took over. Trevithick's reputation was high and, though Bolivar had no use for steam engines, he was short of firearms. He conscripted the engineer into the army to design guns for his troops.

Trevithick had been promised a share of the bullion for his work at the mines, which had amounted to over £20,000, but he was never to see an ounce of it. So, once Bolivar had released him, he had to set about restoring his fortunes by whatever means he could. One successful enterprise involved using the salvage techniques he had first tried at Margate. He made a good deal of money from that operation, but still hankered after replacing the fortune he had lost at Cerro de Pasco; so he invested in a gold mine in the Cordillera of Costa Rica. The mine prospered, but he realised that, in order to sell the gold at a good price, he needed to get it to a port. To do that he would have to make the journey himself through a region where no road existed in order to find the best route.

Trevithick set off with his partner, James Gerard and, bizarrely, two school-age boys, who were to be educated in England. They had troubles all the way, culminating in an accident in which Trevithick fell out of a boat and was about to be attacked by an alligator when fortuitously it was shot by an English officer who happened to be walking on the bank at the time. He eventually arrived at Cartagena, having run out of funds and in a desperate condition. By an extraordinary coincidence another English engineer who had also been working in South America was there at the same time. He was Robert Stephenson, who was about to return home to help his father in the construction of a railway and to design a locomotive. He loaned the elder engineer enough money to return home.

Trevithick reached Cornwall in October 1827, eleven years after he had left for his supposedly short visit. He had hoped that the money made on the plunger pole invention would have provided his family with an income, but it had not been the great success he had predicted. Jane had been helped out by her family, who had put her in charge of a hotel at Hayle intended for the use of important visitors to the Harvey works. It must have been difficult for her to adjust to having him back and a problem for him to try and find some way of restoring the family fortunes. He never stopped trying out new ideas, some of which did make money, including a portable room heater. He was also planning a new type of boiler, specifically designed for use in steamships. He needed finance to build a prototype but, by this time, the Harvey family had lost all interest in helping him out. He found a backer in John Hall, who had an engineering works in Dartford, Kent. Trevithick went down there in 1832 but, before any new ideas could be developed, he died on 22 April the following year and was buried in the local churchyard. Early accounts state that he was buried in an unmarked pauper's grave. Research in Dartford, however, happily produced accounts of the actual funeral and of visits to see the headstone of the great man.

Trevithick had an extraordinarily inventive mind and, had it been married to a good business sense, he would have died a wealthy man. That was not the case. Alternatively, if he had joined forces with someone who could have looked after those matters, he would have succeeded, but that only happened too late in his life to make a difference. His greatest achievement was undoubtedly the development of high-pressure steam engines and ultimately building the first railway locomotives – achievements that others would build on and make the fortunes that eluded their originator.

8

GEORGE AND ROBERT STEPHENSON

The two Stephensons, father and son, are treated together here simply because their professional as well as their personal lives were closely intertwined.

George Stephenson's father, Robert, was a fireman at Wylam colliery in Northumberland, whose job was to feed coal into the boilers for two Newcomen pumping engines. Their home was beside the tramway that carried coals from the pit to the Tyne near Newcastle. The rails have long since gone, but the track is now a footpath and the house is still there. It looks quite spacious and comfortable, but the Stephensons had to share it with three other families and it must have been a noisy place, with an estimated 14,000 chaldrons, the local coal wagons, trundling past every year. It was here that George was born on 9 June 1781.

When he was 8 years old, the family moved to nearby Dewley Burn colliery and a cottage close to the mine. It also marked a turning point in

Images: George and Robert Stephenson. (From Smiles, *Lives of the Engineers*, 1862)

the boy's life. His very rudimentary education had come to an end and he went to work on a local farm, looking after a small herd of cows for two-pence a day. He was soon given more jobs around the farm and his wages were doubled, but he was not an enthusiastic farm hand and, as soon as he was old enough, he joined his brother James at the colliery as a picker – a job that is just what the name suggests. As coal came up from the pit, he had to remove all the rubbish that was mixed in with it.

This was a rather chaotic time with pits running out of profitable coal seams and being closed, while new ones opened. The family were soon on the move again to another colliery near Newburn and another crowded cottage. By the time George was 15 years old he was promoted to the post of fireman at the Mid Mill mine, starting the job with another boy. As the pumps had to be kept going twenty-four hours a day, each had to work a twelve-hour shift. George was gathering experience and, when he reached the point where his wage reached the giddy heights of 14s a week, he declared himself a 'made man'. He was still in his teens. Another family move took them to Water Row, near Newburn and, while Robert was still a fireman, his son George had a new job, as the plugman – the 17-year-old now had a more responsible job than his father. He had a vital role in the smooth running of the engine. A common problem came when water levels dropped and holes in the pump barrel started sucking in air instead of water. George would have to go down the shaft and plug the holes, hence the name.

He became interested in the working of the engine itself and would have liked to learn more, but all the information was written down and he could not yet read. He decided to do something about that by enrolling for lessons in reading and writing, for which he paid 3d a week; when a night school opened nearer his home at Newburn he enrolled, even though it cost him an extra penny. He was keen to learn and persuaded the schoolmaster to give him extra exercises to do at home. He was never a confident letter writer, but his reading skills were now sufficient for him to be able to read what others had written and learn from them. These studies, combined with his opportunity to make practical observations on how steam engines worked, set him on the path to becoming a successful mechanical engineer.

Stephenson was given an even more important job in 1801 when he was appointed brakeman at the Dolly Pit at Black Callerton. He was in charge of a whim engine, which was used to raise and lower men and material in the shaft. The brakeman had to rely on indicators above ground to give the position of the cage in the shaft, and it was down to his skill to ensure that everything went smoothly and that the cage stopped exactly where it should.

George was now earning £1 a week and was making extra cash by repairing shoes. He took lodgings at a farm and met a local farmer's daughter, Frances (Fanny) Henderson. She was then 31 years old, but that did not stop the young man falling for her; when he felt he had enough money set aside he proposed and they were married in 1802. He was now offered a new job as brakeman at Willington Quay on the Tyne almost opposite Jarrow. The ships that arrived to fill with coal rarely had anything to bring to Newcastle, so they travelled with ballast that had to be removed and piled up behind the quay. It was at the foot of this great heap of stone that the young married pair took a room in a cottage.

One day while Stephenson was at work there was a fire in the cottage chimney, which was dowsed, but everything was soaked and covered in soot. Among the casualties was an eight-day clock, a much-prized possession. Stephenson decided to mend it himself, succeeded and found a new way of making extra money as a clock repairer. In October 1803, Fanny gave birth to a son, Robert, and the following year George was offered post as brakeman at Killingworth and the family moved there. In July 1805, Fanny gave birth to a daughter, also named Frances, but the child died when just three weeks old. Fanny herself became ill and she too died in May the following year. Mother and daughter were buried side by side at Long Benton. For George Stephenson the double tragedy may have been the motive for leaving the area altogether, which had such unhappy associations, and he moved to Scotland to take charge of a Boulton & Watt engine at a textile mill at Melrose, leaving Robert in the charge of a housekeeper.

George was not happy in Scotland and, after just a year, he left to walk back to Northumbria. He had been frugal and saved £28, but when he arrived back, he found that his father had suffered a severe accident and was no longer able to work. With no income, he had run up debts of £15, which now had to be paid off. At least there was enough money for George to be able to take on a cottage at West Moor, near Killingworth. Here he was able to look after his father and young son. At first, he employed a local woman to look after the family while he was at work, but when that proved unsatisfactory his unmarried sister, Eleanor, moved in. The new home was known as Dial Cottage, from the sundial above the front door, and it was to be his home for many years. Over the years, he extended it to make a far more comfortable home than he had ever known before. He was now back at his old job of brakeman, but everything was to change thanks to an old Newcomen engine at the West Moor pit.

The engine was no longer able to raise the water efficiently, and there was a real danger of the mine being flooded. Stephenson felt he had the answer, and the head viewer Ralph Dodds let him have a go at solving the problem. George had recognised that the steam was not condensing, mainly because the nozzle of the water jet was too small and the water pressure too low. He put on a new nozzle, raised the water tank a further 10ft and then increased the steam pressure from 5psi to 10psi. When it started up again it was alarmingly active, but soon settled down to a regular beat and within two days all the water had been cleared.

The following year, the Killingworth engine wright was killed in an accident and, thanks to the good impression George had made, he was offered the post. He was not just looking after the Killingworth engines, but was now in charge of every engine in the Grand Allies colliery group, the most important mines in the whole of north-east England. George was now on a salary of £100 a year and had an agreement that allowed him to work for other concerns when appropriate. It was not so much a step forward in his career, but a huge leap. He soon became busy making improvements, including extending the tramway network and installing an underground rope haulage system at Killingworth. He now had enough money to pay for a good education for his son, who at the age of 12 was sent to a highly respected school run by Dr Bruce in Newcastle. This was 10 miles away, so Robert was given a donkey to take him there and back. The boy proved to be a good scholar, but must have found it difficult mixing with the smartly dressed boys from middle-class families when he was a roughly clothed miner's son. But eventually he settled in and, unlike his father, lost his Geordie accent.

The next major change in George Stephenson's life came about as a result of a tragedy. On the morning of 25 May 1812, the ground around the Felling colliery near Gateshead was shaken by an immense explosion at the mine. Eighty-seven men and boys were trapped underground but, in spite of all efforts, it proved impossible to reach them. How many died in the explosion and how many simply died from hunger will never be known. It is almost certain that the explosion was caused by an escape of methane ignited by the naked lights used by the miners. A committee was set up to try and find a safe way of lighting the mines. They invited one of the leading scientists of the day, Sir Humphry Davy, to investigate and, after extensive experiments in his laboratory, he came up with the famous safety lamp that has remained in use in deep mines ever since. He was awarded £2,000 for this lifesaving device.

Davy was not, however, the only one eager to solve the problem. George had no laboratory to work in, so he had to make his experiments below ground. He bravely went to a spot underground where there was a known blower – a slight methane leakage. He held a candle flame to the

blower and observed how it behaved. He designed a lamp on a quite different principle from Davy's. Air was only admitted into the bottom of the glass cylinder through a narrow tube. If methane came in, it forced out the air and, deprived of oxygen, the flame died. This was a sign for the miners to escape.

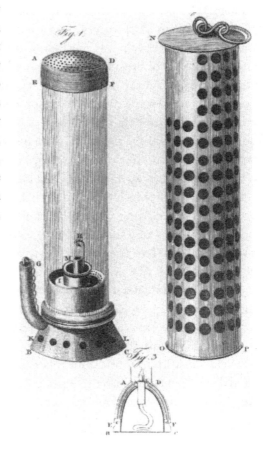

8.1 George Stephenson's safety lamp. (From Smiles, *Lives of the Engineers*, 1862)

It worked and was used in the north-east for many years, where it was known as the Geordie lamp. Locals felt that, as Davy had been paid for his invention, so too should George receive his own award and a subscription was raised. Davy was incensed: to him it was inconceivable that a common workman could be considered as capable of producing a lamp when he lacked proper scientific training, and he denounced it as a fraud. George was equally furious at being denigrated in public and it bred in him a distrust of alleged experts, and London experts in particular, that was to stay with him all his life.

The Napoleonic wars had brought many difficulties to the country and one of them was the rapidly increasing cost of fodder for horses. John Blenkinsop, of the Middleton Colliery just outside Leeds, needed to send his coal on the horse-drawn tramway to the Aire & Calder Navigation for shipment. He decided that the time had come to reinvestigate the possibility of using steam locomotives. Trevithick was, of course, still in South America, but Blenkinsop was able to buy out his patent. He was still faced with the problem of breaking rails. He could use a lighter engine, but that would not have had the power to haul the trucks. What he needed was a way of increasing traction. Blenkinsop decided that the answer was a rack and pinion system in which a cog on the engine would engage with a toothed rail. Ideally the rail would have been set between the rails but, as the tramway was still likely to be used by horses, he had to use the rack rail to one side of the track. He turned to a local engineer, Matthew Murray, who had experience in building stationary steam engines, to design the locomotive. There were inevitably many similarities to the Trevithick engine but, instead of a single cylinder, there were now two set vertically in the boiler, with two connecting rods geared to a central shaft on which the cog was set. Unlike the Trevithick engine, the boiler was just a single, straight flue and exhaust steam simply blew out into the atmosphere. Two engines were built, *Prince Regent* and *Salamanca*, and they began work in August 1812. The experiment was a success and the engines were able to haul loads of 94 tons at a steady 5mph.

Several illustrious visitors came to see the engines at work, including a grand duke from Russia, later to become Czar Nicholas I, which probably explains why Russia was among the first countries to invest in steam railways once they had been established in Britain. Another visitor, who probably did not receive the formal welcome bestowed on the grand duke,

8.2 The Middleton Colliery Railway, which George Stephenson visited, and the engine on which he modelled his own first locomotive. (Mansell Collection)

was George Stephenson. He was sufficiently impressed to make extensive notes on all aspects of the mechanism. Other engineers also arrived and inspected the railway, and some began producing their own designs without including the rack railway system.

Among the first were William and Edward Chapman, who developed a system in which the engine hauled itself along a fixed chain by means of a grooved wheel on the locomotive, which seems to have been mainly intended for use on steep slopes. They then built a more conventional engine, which reduced the risk of breaking the track by mounting it on two four-wheeled bogies. All eight wheels linked by gearing, which must have severely reduced the available power. The most important advance was made by the Wylam colliery engineer, William Hadley. His first attempt with a single-flue boiler, a single cylinder and a flywheel was not a success. The next version had a return-flue boiler and two cylinders, set vertically to each side of the boiler. The drive system was complex, using a pivoted beam, connecting rods and gears. Hadley had not, however, solved the track breakage problem so, for subsequent engines, he borrowed the Chapman idea of using two bogies. Four engines were built, two of which, *Wylam Dilly* and *Puffing Billy*, survive but in a later adaptation with just four wheels. There was one other bizarre engine, designed by William Brunton, which dispensed with wheels altogether and moved using steam-powered legs. It was unlikely to have worked, but we shall never know how it might have performed as it blew up at the start of the first trial, killing the crew.

It was obvious that the most important mine owners in the region, the Grand Allies, would not want to be left behind in the race to develop a successful locomotive and it was equally inevitable that they would turn to Stephenson to design and build one. It is clear that his first effort depended very much on what he had seen at Middleton. It too had a pair of vertical cylinders set in the boiler, a straight flue and exhausted to atmosphere. It was first steamed in 1814 and was named *Blucher* after the Prussian general who had led his forces against Napoleon. It was driven through gears, which caused very uneven running, and when they became worn they rattled alarmingly. It was unreliable and, on one occasion, came to a halt right across a main road. George's brother, James, was driving at the time and, as it was quite near his cottage, he went and fetched his wife to help push it clear. George began to look for ways to improve its efficiency. One idea was to send exhaust steam up the chimney as Trevithick had done to increase the air flow through the firebox, but with such a primitive boiler that would have had little effect. He then made more radical changes, attaching connecting rods from the pistons to crank pins cast into the wheels. With two pistons working independently, he had to set the motion at 90 degrees to maintain momentum. He then added another variation, linking the two pairs of wheels by continuous chains, similar to the familiar bicycle chains of today. So far nothing he had done had represented any real advance on what others were doing at the time. There was, however, one difference. Where the other engines all had the familiar right-angled plates for rails, the Killingworth route had been laid with edge rails. *Blucher* was the first locomotive to be fitted with flanged wheels.

All these early locomotives had to deal with the problem created by the rails on which they ran. One of the enthusiastic backers of the Geordie lamp had been William Losh of the Losh, Wilson & Bell ironworks in Newcastle. In 1815 he invited to George to work with him for two days a week at a salary of £100, and the Grand Allies agreed to the arrangement. They began by looking at ways in which they could minimise track

8.3 An early Killingworth locomotive with chain drive. (Science and Society)

damage. Chapman had patented the idea of the bogie, so George looked for an alternative and the solution he came up with was the 'steam spring'. The wheel axles were held in bearings attached to piston rods from cylinders set into the boiler. As steam pressure rose it had a cushioning effect. The device remained in use until sufficiently substantial metal springs were introduced. George's early locomotives were four-wheelers, but in 1817 he supplied a six-wheel version for the Kilmarnock & Troon plateway in which all six wheels were linked by chains.

He and Losh next turned their attention to improving the smooth running of engines. All the tracks then in use were mounted on stone sleeper blocks, with the rails abutting end on. The stone blocks were liable to move, causing the rails to come out of alignment. The partners designed rails that met in lap joints, so that seen from the side the join was no longer a straight line but in an S shape. These were mounted on chairs with a slight curve so that, even if the block slipped, the rails would slide along the curve and remain attached and level. The whole Killingworth track was relaid with the new rails.

George was equally busy with his familiar work around the mines. One recurring problem arose from one of his own developments. He had devised a ventilation system that used flues that passed from the surface down to the lowest level of the mine. There was always the danger that soot accumulating in the flues might catch fire and even set fire to the coal levels through which the flue passed, which it did on a few occasions. Whenever there was trouble, George was always the first to go down the mine to deal with it, but he ran the risk of being overcome by what the miners called choke damp: carbon dioxide. Robert recorded in a postscript to a letter that his father had been overcome by choke damp on two or three occasions.

George was unique among the early pioneers that he not only continued to work on developing the locomotive, but was equally aware that success depended on having good track. He saw the system as a whole. He was not, however, the only enthusiastic supporter of railways. William James was a land agent in Henley-in-Arden who, at the height of the Napoleonic wars, wrote to the Prince Regent proposing a tramway to link the major dockyards of Chatham and Portsmouth, and even more ambitiously proposed a line from Stratford-upon-Avon to London. Neither scheme went forward. He did, however, visit Northumbria and was impressed by the Stephenson–Losh rail, offering to become their agent for selling it in all areas south of the Humber. He would later play a far more important part in the Stephenson story.

In 1819, the Hetton Coal Company was formed to open up a mine near Hetton-le-Hole in Durham. The coal would need to be exported from Sunderland, so the company needed a transport route to the River Wear. George surveyed the route and faced one major obstacle, a line of hills at Warden Law that lay between the mine and the river at Warden Law. He decided to construct two inclines. Two steam engines were installed to haul the loaded trucks up to the summit, but on the northern slope the incline could be made self-acting – the weight of the loaded trucks going down being used to pull up the empties. George supplied five Killingworth-type locomotives for use on the line, which was laid with the Losh–Stephenson rails.

At this period Robert had finished school and was apprenticed to Nicholas Wood to get a training in mining and transport. It was almost the end of his career. He had gone with Wood and the under viewer, Moodie, to inspect a rock fall. Wood clambered up onto the pile, carrying a lit candle. Gas had accumulated above the rocks and there was an explosion. Robert and Moodie ran back towards the shaft, but realised that Wood was not with them. They rushed back and found him stunned, but with only minor injuries. They all made it safely back to the surface. After that Robert was employed for a time at West Moor colliery.

The biggest change in the fortunes of the Stephenson family came with the development of the Stockton & Darlington Railway. The story begins in 1810, when a meeting was held to discuss ways of improving transport from the south Durham coalfield to the Tees at Stockton. The question was raised as to whether it should be a canal or a tramway. A committee was formed and, after two years of dilatory discussions, this authorised the tramway option. There now followed a survey of the route and that took an equally lethargic three years to complete. By then the country was in a financial crisis and there was no money available. The subject was raised again in 1818 and a new survey was ordered, but this time for a canal. When the report came in, the route bypassed Darlington and was rejected. The engineer George Overton was now brought in, an experienced builder of tramways in South Wales, who demolished the arguments of the canal faction by producing figures showing that the tramway owners in that region were receiving dividends of 7 per cent while the canal owners were only getting 1 per cent. The line was surveyed and, in spite of strong opposition from powerful landowners in the area, an Act was passed in 1821 for a conventional plateway to be worked by horses, though there was provision for stationary engines to work steep slopes.

It was at this point that George Stephenson and Nicholas Wood went to see Edward Pease, a prominent Quaker businessman in Darlington and enthusiastic supporter of improving transport in the area. There are many different stories about that meeting, most of which suggest that the two men were humble workers arriving unannounced to speak to the great man. In one version, they were shown into the kitchen, in another they walked barefoot from Newcastle and so on. But George was a considerable figure, already made famous from his invention of the safety lamp and, although he was called many things over the years, no one ever referred to him as humble. Fortunately, we have Nicholas Wood's own account of what really happened. The two men had made an appointment to meet Pease and had travelled on horseback to Newcastle, from where they took a coach to Stockton-on-Tees. They did walk to Darlington, but only so that they could become familiar with the route of the proposed railway. The meeting was a success, and Pease was persuaded that what was needed was not a plateway, but a track to be laid with edge rails and worked by steam locomotives. A second Act of 1823 changed the whole nature of the line, which instead of allowing for a private railway specified that it would be a public railway that was authorised to carry both goods and passengers, and that it was 'empowered to erect steam-engines on or near the railway, and also to make and use locomotive engines thereon'. This is the significance of the line: the first public railway in the world authorised to use steam locomotives.

George was invited to be in charge of the survey, but he was a busy man, still being engaged on work on the Hetton line, and he needed to settle just what he would be required to do and how he was to be paid. There were to be further meetings with Pease and in July 1821 George reached agreement to begin the survey. He asked Pease to recommend an assistant who knew the country well, and he recommended George Dixon. He also appointed Robert as an assistant surveyor. As work got started news reached him of a new type of rail, invented by John Birkinshaw at the Bedlington iron works. He used a system for rolling wrought iron to produce a rail, roughly 1in wide and 5in deep and available in 20ft sections: later versions were fish-bellied, the lower part of the rail being curved. The wrought iron rails were more expensive than cast iron, but were so obviously superior that it was eventually decided that they should be used for the new line. Losh was furious at the loss of such a lucrative contract and felt he had been betrayed by George. Relationships between the two men never recovered.

In spite of its name, Darlington was never the terminus; the line extended beyond that to a colliery near Bishop Auckland. As with the Hetton tramway, inclines were constructed where the route was crossed by two ridges, at Bressington and Etherley. The former was the more impressive, with a total rise of 150ft that originally used one winding drum worked by a 30hp steam engine. In practice this proved to cause delays, so a second winding drum was added. Elsewhere, George used the technique first employed on the canals of cut and fill to create cuttings and banks. There were two major bridges on the route: one over the Skerne at Darlington was a conventional three-arched stone bridge, designed by Ignatius Bonomi; the other across the Gauntless was designed by George himself. It had a 50ft span and was constructed of wrought iron girders on cast iron supports, not always an ideal combination as his son would later discover.

Once the survey was completed and work could get under way, Robert's job as assistant surveyor was over, but instead of moving to another job he was enrolled at Edinburgh University. He studied natural philosophy – science in today's language – and became interested in the comparatively new subject of geology. However, he only stayed for six months, probably because he found there was little he was learning that was likely to help him develop as a practical engineer. Meanwhile, as work went on with constructing the line, it became obvious that no one apart from George was now showing very much interest in building locomotives to run on it. If the line was the success that the Stephensons believed it would be then there would be a demand for more similar lines and more engines to run on them. A new manufacturing company was formed to be called Robert Stephenson & Co., financed by shareholders: the two Stephensons, Pease and Michael Longridge of the Bedlington iron works. Pease loaned Robert the £500 he needed for shares, and the 19-year-old was now nominally put in charge of the works at a salary of £200. His conditions of employment clearly specified that in fact George would be in charge of all the planning and designing. It seems an odd arrangement, but it was probably felt that some awkward questions might be asked if George Stephenson, the company's engineer, was ordering locomotives from himself.

8.4 A replica of *Locomotion* at the Beamish Open Air Museum. (Beamish Open Air Museum)

One of the first jobs for the new works when they opened at Forth Street in Newcastle was to build locomotives for the Stockton & Darlington. The first two, *Locomotion* and *Hope*, had several differences from the Killingworth-type engines. The biggest change came with the drive, which now used connecting rods and dispensed with the chains. There were two cylinders set in line in the boiler, above which were crossheads, with the motion controlled, rather like that in a beam engine, but pivoted at one end rather than in the centre. The connecting rods were attached to crank pins, set at a quarter turn from each other. The engines still had primitive single flue boilers, with the firebox at one end, and the far end turned up to create the chimney. There was no reversing gear so, in order to make the engine go in the opposite direction, the driver, who stood on a platform at the side, had to unclip the connecting rods and reattach them when the cranks were in the right position. Watching this operation in practice on the replica at the Beamish Open Air Museum makes it seem both difficult and rather dangerous. The two engines were adequate for the job in hand, trundling along with loaded wagons at what was at best a brisk walking pace. With such machines, there was no possibility of attracting passengers, but a service was set up using a conventional stage coach *Experiment*, fitted with flanged wheels and pulled by horses. The journey between Stockton and Darlington took two and a half hours. A maintenance works was established at Shildon, with Timothy Hackworth in charge.

The Stockton & Darlington was not the only new line being proposed at this time, and William James was foremost among the promoters. He had not succeeded in getting a railway from Stratford to London, but he did succeed in promoting a tramway from that town to Moreton-in-Marsh: the bridge that carried it over the Avon at Stratford still survives. He now worked on another ambitious scheme, a railway to connect Liverpool to Manchester. Since the middle of the eighteenth century, the cotton industry had boomed, and cotton from the United States would arrive at the port for shipment on to the heart of the Lancashire industry. The main transport route, however, was still the Bridgewater Canal. According to the railway promoters, it was often the case that the cotton took less time to come from America to Liverpool than it did to get from there to Manchester. James approached a wealthy Liverpool merchant, Joseph Sanders, for support, and he in turn got the support of two local MPs, George Canning and William Huskisson.

James agreed to make a survey of the line and was joined by his brother-in-law, Paul Padley. Robert Stephenson joined the team, which began work in 1822. It soon became apparent that there were formidable difficulties in their way, not least how to cross the morass of Chat Moss. They also faced opposition from communities along the way. Landowners, many of whom had shares in the canal, refused the surveyors access to their land, while villagers often hurled insults and more solid missiles at the surveyors. James, who should have been leading the team, was frequently absent. He was an enthusiast who was never seemingly content to keep just one project on the go. He was now busy looking at a line to link Canterbury and Whitstable. Unfortunately, he was neglecting his own business and was declared bankrupt and sent to the debtors' prison in Southwark. The promoters of the London & Manchester Railway (L&MR) and the Canterbury & Whitstable now contacted George Stephenson, who was appointed chief engineer for both lines. James lost all connections with the schemes he had so ardently promoted and he spoke bitterly of Stephenson's 'duplicity'.

Both father and son were busy on a variety of projects, with George being consulted on a variety of railway projects and Robert occupied with the Forth Street works, when an offer came through from one of the partners who had interests in mining projects in South America. He wanted someone to visit the mines in Mexico to investigate the possibility of building railway connections to the coast. George's brother, Robert, was first consulted, but his wife was utterly opposed to the idea, so instead the younger Robert was offered the job. There has been much speculation on why he accepted the offer. L.T.C. Rolt, in his biography of the Stephensons, suggests there might have been a rift between father and son, but it seems equally feasible that Robert simply liked the whole idea. Though nominally in charge at Forth Street, all important decisions were taken by his father and he was now being offered the chance to visit what would have seemed an amazingly exotic country. What young man could resist the temptation? His father objected, but Robert promised he would not be away for long and would return as soon as he had surveyed the possible routes; so off he went.

George, meanwhile, had to take charge of the survey for the L&MR. He was still busy on numerous projects, so much of the work had to be left to assistants. There were still the same problems with landowners, and one of them called Bradshaw arranged to have guns fired through the night to prevent surveyors sneaking in after dark, while Lord Sefton employed a small army of gamekeepers to protect the ground. The surveyors got around the latter by having guns fired in one part of the ground, causing the gamekeepers to rush over in the hope of catching poachers, while the surveyors sneaked in. It was not conducive to preparing an accurate survey. As a result, when the Bill came up before the select committee of the House of Commons, its defects were exploited by the lawyer Edward Hall Anderson, employed by the opposition. As George Stephenson was questioned, it became more and more clear that he had not made himself fully aware of what was going on and had overlooked several major errors. Anderson summoned up by saying, 'I say he never had a plan – I believe he never had one – I do not believe he is capable of making one.' Stephenson later said he wished he had been able to crawl away into a hole at the end of the session. The Bill was thrown out: it was a day of complete humiliation. He lost his position as chief engineer and had been publicly mocked.

Things were at least going well on the Stockton & Darlington, and six months after the Parliamentary debacle the grand opening of the line was set for 27 September 1825. There was to be a procession consisting of a train drawn by *Locomotion* drawing six wagon loads of coal, followed by the proprietors in the coach *Experiment*, six wagons fitted with seats for the gentry and another fourteen wagons with standing room only for everyone else. The workers who had built the line followed on behind in horse-drawn wagons. The whole procession was led by a horseman carrying a banner. Crowds turned out to watch and, when the train arrived at Stockton, it was greeted by a band playing the national anthem. After that there was a formal dinner with toasts, the last of which was to George Stephenson on his day of triumph.

Meanwhile, the L&MR had to find a new engineer to resurvey the route for a fresh attempt to get a Bill passed. They appointed George Rennie, and he asked for an 'operative engineer' to work under him. Two suggestions were made by a company committee – John Urpeth Rastrick, an experienced engineer, and Stephenson. Rennie rejected both – and certainly one could not imagine Stephenson working as an assistant to Rennie. The latter remarked that he would be quite happy for someone like Telford to work for him – an extraordinary piece of arrogance given Telford's status. Rennie did, however, recommend a young engineer,

Charles Vignoles, who had been an army officer who had left at the end of the Napoleonic wars. But when it came to settling who should be in charge of operations, Rennie was unable to agree terms. They then appointed Josias Jessop, son of the canal engineer, as 'consultant engineer' and invited Stephenson to take over as 'working engineer'. That did not work out well. The two men disagreed over the line to be taken and, although Jessop was nominally in a superior position, the committee always agreed with Stephenson. It was an impossible position and Jessop resigned. Stephenson was once again the chief engineer.

During all this period, Robert had made his way to South America. He had been forced to spend some time in London before setting off, but used it well, visiting the Royal Mint, which gave him a new idea for a way of striking coins, and attending lectures on mineralogy. He eventually arrived at La Guaira in what is now Venezuela, but was then Gran Colombia. He disembarked with his assistant, Charles Empsom, while the remainder of his party sailed on to Cartagena to make their way to the mining area and start installing machinery. Robert's first task was to consider whether a breakwater could be built at La Guaira and whether it would be possible to build a railway from there to the mines. He decided a breakwater would be too expensive, but recommended that a pier could be built, so that vessels could unload directly, rather than having to anchor in the bay. One look at the hills surrounding the town was enough to convince him a railway was impractical.

Robert's next task was to assess the mines of Colombia. He spent four months making his way on difficult tracks to the capital, Bogotá, at 8,800ft above sea level. From there he eventually made his way to the head of the navigable Magdalena River at Honda, where he found the rest of the party, but with all the heavy machinery still unpacked on the river bank. It turned out that the final 12 miles to the Santa Ana silver mines were up a narrow mule track climbing the steep slope of the Andes. Robert reported back to London but, by the time his message arrived, another shipment was already on its way. Robert, however, made his way to the mine, where he had a comfortable house and from which he set off on more expeditions to survey the land for suitable sites for development. Throughout the time he spent at the mine, he had trouble with the miners, who had arrived from Cornwall and who were stuck out in the wilds with little to spend their wages on apart from drink. As a result, Stephenson complained he could scarcely get more than half a day's work from anyone. He stayed on, but as his contract came near its end, he received urgent messages from England that he was badly needed back home. As we know, he made his way to Cartagena and a meeting with Trevithick. It would be fascinating to know what they talked about: the older man, who had invented the locomotive, would now be hearing for the first time that, in his absence, steam railways were being run and more promoted throughout Britain. Robert travelled first to New York, which did not impress him, and briefly visited Canada, where he felt more at home. He arrived back in Liverpool in November 1827.

George Stephenson still had many projects on his hands beside the L & MR. He was engineer for the Bolton & Leigh line and was being consulted about a possible railway linking Birmingham and Manchester. He was also responsible for the Canterbury & Whitstable, but now he was able to hand that work over to Robert, who was delighted since, as he wrote to a friend, 'it is closer to London'. The attraction there was a young lady he had met before leaving for South America, Fanny Sanderson – they would eventually marry in 1829 and make their home in Newcastle.

The proposed L&MR line involved considerable engineering works. At the Liverpool end, there were to be three tunnels and a deep cutting blasted out of the sandstone at Olive Mount. There was an immense via-duct over the Sankey Navigation and a smaller one over the Bridgewater Canal. There were embankments to be built up to a mile long and cut-tings, of which the longest was near Newton-le-Willows, and sixty-three smaller bridges. The greatest obstacle was still the crossing of Chat Moss. Construction got under way and, when the first tunnel was started, the pilot bore was found to be 13ft off line and threatened to undermine houses. It had to be abandoned and a new bore started in the correct place. The engineer in charge of that end was Vignoles, and Stephenson blamed him for the mistake. Relations between the two were never good and Vignoles resigned. The two new engineers working under George were his son, Robert, and a former Forth Street apprentice, Joseph Locke.

When test borings were made at Chat Moss, the workers had to go down 14ft before hitting solid ground. The first attempts to drain the morass were a failure: as fast as trenches were dug, the bog simply oozed back in again. This problem was solved by constructing drains out of rows of barrels with the ends knocked out. Then the navvies started piling on

8.5 The Liverpool & Manchester line across bleak Chat Moss. (From I. Shaw, *View of the Most Interesting Scenery on the Line of the Liverpool and Manchester Railway*, 1831)

earth to create a foundation for a low embankment, but it never seemed to appear above the surface. Stephenson tried a different approach, floating rafts of brushwood and heather onto which the earth was piled, reasoning that an equilibrium must inevitably be reached. That worked. At the Manchester end, however, the piling of earth seemed interminable. Eventually, however, a bank was created and a single line laid on top. Work on the tunnels was now going ahead with Locke in charge. The Edgehill tunnel was on an unprecedented scale: 16ft high, 22ft wide and a mile and a quarter long. In all 300 men were employed in the works. Amazingly, visitors were invited in to see the work in progress.

Although work was going ahead on the line, one vital decision had not yet been taken. How was the line to be worked? One faction favoured locomotives, while another would have preferred a very different system that would involve a series of stationary steam engines set up along the line, with trains hauled from one to another by cable. Two experts, Rastrick and James Walker, were appointed to travel the country to look at the different ways of working and to report back. In the meantime,

Robert Stephenson and Joseph Locke were given the task of writing a pamphlet to make the case for the locomotive. One crucial argument was that it only needed something to go wrong with one of the stationary engines for everything to come to a halt until it was repaired, while a locomotive could simply be taken out of service and trains could still run. The final report, however, was equivocal. One thing against the locomotives was that the ones already in use were slow and cumbersome. Their supporters claimed they were capable of great improvement, but could not yet claim to have any support for their argument.

The committee decided to put the whole question to the test. They would mount a trial on a level section of the route at Rainhill and set out what an engine would have to do to pass the test. The rules were slightly different for different weights of engine. For the largest allowed, with a weight of 8 tons, it had to be able to cover an equivalent distance of a return trip along the whole route, with a 20-ton train of carriages at a speed of 10mph at a boiler pressure of no more than 50psi. The prize was £500 over and above the cost price, but the successful builder, if any were successful, could be assured of valuable contracts in the future.

Many entrants came forward with ideas, mostly more fanciful than realistic, but there were a few serious competitors, including the Stephenson works at Newcastle. Robert, back from South America, was full of new ideas. One idea was put into practice when an engine was ordered to help with construction work at Liverpool. He placed the cylinders outside the boiler and angled them at roughly 45 degrees with a connecting rod to a crank on the front wheels. The engine, the *Lancashire Witch*, was later used on the Bolton & Leigh Railway. This was the starting point for what was known at the time as the 'Premium Engine'. The next important idea came from the L&MR company secretary, Henry Booth. He suggested that steam could be raised much more efficiently if, instead of passing through a single flue, whether doubled back or not, the hot gases from the fire were to pass through many small tubes set in the boiler. There was one other element needed to make the engine more efficient, a good heat from the fire. Here Robert returned to a device first used by Trevithick: the exhaust steam from the cylinders, instead of blowing away into the air, was allowed to escape through a narrow tube at the bottom of the chimney. This resulted in air being drawn through the firebox. The three elements had been decided, and working drawings were made and construction began. One other novel

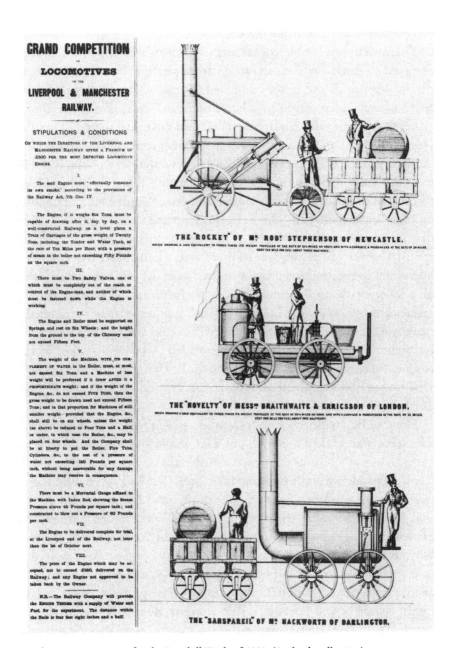

8.6 The main contestants for the Rainhill Trials of 1829. (Author's collection)

feature was a reversing gear, operated from the footplate. By the time it was completed it had a new name, *Rocket*.

There were few viable contestants. Timothy Burstall of Edinburgh designed a vehicle based on an earlier steam carriage, but it was damaged on the way south and never competed. There was an oddity that consisted of a machine driven by a horse on board walking on a treadmill, which hardly met the requirements set and never competed. Timothy Hackworth of Shildon entered a conventional return flue engine, *Sans Pareil*. A replica built to celebrate the 150th anniversary of the Rainhill trials showed it to be a sound engine, but nothing about it was capable of further development. The public favourite was an engine built in London by John Braithwaite and John Ericsson, who had developed the world's first steam-powered fire engine. They decided to go for lightness and speed. The vertical piston was connected to a bell crank, from which a horizontal connecting rod drove a cranked axle, the first use of this device on a locomotive. In the centre of the platform was what was basically a vertical stove, with the boiler itself being pipes wrapped around the stove and the convoluted exhaust pipe. The stove was closed by a lid and had to be fed from the top. Blast was provided by bellows worked by the engine and the whole was mounted on spoked wheels. To the spectators who turned up to see the trials it looked sporty and exciting, but George Stephenson had a different view – 'nae guts'.

The trial began officially on 6 October 1829 in front of a large crowd. The track was marked out with poles set a mile and half apart, with an extra eighth of a mile at each end to enable the engines to get up speed. They then had to trundle back and forth until they had covered a distance equivalent to a journey between Manchester and Liverpool. There was then a pause before they had to set off to simulate the return journey. The Hackworth engine was found to be considerably over the weight limit, but was allowed to run anyway. Neither *Novelty* nor *Sans Pareil* were able to complete the course, but on 8 October *Rocket* ran the full distance at an average speed of 14mph. On the last lap, George Stephenson opened up the regulator and the engine shot along at 30mph. It was a total triumph. There were mutterings in the Hackworth camp that their engine had failed because of a faulty cylinder cast at the Newcastle works, but the problem had mainly been caused by a failed feed pump, which had not been made there. The main outcome was that all talk of moving trains along by cable haulage was abandoned, and in future all steam locomotives would rely on the basic elements developed on *Rocket*. That engine was later altered and the cylinders moved closer to the horizontal, as can be seen in the original now owned by the Science Museum in London.

The leading proponent of fixed engines in the company was James Cropper, who was still, it seems, prepared to oppose the Stephenson family. He insisted on the company purchasing two engines based on the *Novelty* design, neither of which proved satisfactory, and they were reduced to the lowly role of contractors' engines before being sold off. He then decided to employ another engineer, William Chapman, to inspect the works. That did not go well either. When Chapman questioned one of the workmen about locomotive performance and was unhappy with the answers he got, he sacked the man on the spot. This was one of the men who had been brought down from Newcastle and George was furious. There was only one possible outcome: Chapman went.

While preparations for the opening of the L&MR were going forward, work on the little Canterbury & Whitstable was completed. A *Rocket*-type engine, *Invicta*, was purchased and the line was officially opened in April 1830. It could claim the honour of being the world's first railway to offer a passenger service operated by steam locomotives, just pipping the L&MR to the post.

Meanwhile, over at Liverpool, once again a tunnel proved to be out of alignment due to mistakes in the survey, but fortunately the errors were discovered in time for corrections to be made. Robert Stephenson had been steadily working on improvements on the locomotives. Being no longer tied to the weight restrictions imposed for the Rainhill trial, he was able to increase the size of the engine, culminating in the construction of *Northumbrian* in which the number of boiler tubes was increased from the twenty-five on *Rocket* to 132. The firebox was no longer separate, but an integral part of the boiler, which had now been enlarged. Cylinder size had been increased and there was now a proper smoke box at the chimney end. Robert was always looking for improvements, and the next locomotive he designed looked very different. *Planet* had an outside frame and the cylinders were now under the firebox and drove to a cranked axle. A new reversing system was also designed for the engine. It was to be the first of the very successful Planet class that served the railway well.

The new line would also require rolling stock, and here the company turned to specialist coach builders for their passenger vehicles. The first class were four-wheeled vehicles that looked rather like three conventional stage coaches that had been stuck together; the second class were open sided, but had an overhead canopy to provide some protection from the weather, while the third class simply had open trucks with benches. The gentry could actually travel in their own private coaches mounted on flat trucks. There was also a need to build a new type of building: a railway station. The first was built at Liverpool Road, Manchester, in the Georgian style and, from the outside, could have been mistaken for a gentleman's residence.

The grand opening was set for 15 September 1830. The principal guest was the Duke of Wellington, who was not a popular choice. He was still remembered in Manchester as the leader of the party that had ordered the troops in to break up a peaceful demonstration to demand the right to vote being given to all citizens. The resulting violence on that day in St Peter's Square became known as the Peterloo Massacre. However, the duke was to have his own elaborate carriage, complete with a gilded canopy topped by a coronet. Eight locomotives were to set off from Liverpool with eminent guests, including the MP who had supported the railway, William Huskisson. There was to be a stop to take on water at Parkhead, and instructions were given to guests not to leave their coaches. However, Huskisson saw the duke across the tracks and went across to speak to him. Other passengers saw *Rocket* steaming at speed towards him. The shouts of 'get in', only seemed to confuse Huskisson, who was hit by the engine and his leg crushed under the wheels. *Northumbrian* was uncoupled and a flat truck attached with the injured man on board. With Stephenson on the footplate, it sped off towards the nearest hospital at Eccles, but it proved impossible to save Huskisson's life. In spite of the tragedy, it was decided to go on with the rest of the day's journeys. The arrival in Manchester was not the triumph it should have been as the trackside echoed to cries of 'Remember Peterloo'. With all the changes that had to be made, the final train only arrived back at Liverpool at eleven o'clock that night. It was not the day that anyone had hoped for and expected. It was, however, the world's first inter-city passenger railway, and many more were soon to follow, in some of which George Stephenson himself was a shareholder.

8.7 The viaduct crossing the Sankey Brook Navigation: the canal no longer exists. (From T.T. Bury, *Colourful Views on the Liverpool and Manchester Railway*, 1831)

One of the new lines was the Leicester & Swannington, a short but important route intended to link the coalfields of Swannington to the city. George was already busy on other lines, so the job of chief engineer went to Robert. In the course of the work, he began designing new engines for the line. The first engines were four wheeled, with both sets of wheels coupled in an 0-4-0 wheel arrangement. He found these to be unstable, so he added an extra pair of undriven wheels at the back to create a 0-4-2. He was so pleased with the result that he abandoned four-wheeled engines forever. It was still underpowered, so the next engine had all wheels coupled 0-6-0. By 1833 Robert had a third variant ready. He decided to leave the drive wheels without flanges and introduced the first of what became known as Patentees with a 2-2-2 arrangement. Other new inventions of the time were more efficient steam brakes and the steam whistle.

Robert wrote to his father to tell him that he was convinced there were substantial unworked reserves of coal in the area. George joined with two partners and bought land at Snibston in Leicestershire and began sinking pits. To keep an eye on the operations he moved to a handsome house,

Alton Grange at nearby Ashby-de-la-Zouch. In 1838 he was to move again to a Georgian mansion, Tapton House at Chesterfield.

Rather surprisingly, one new line was built that was designed to be worked by horses. It was a single-track linking Whitby to Pickering, together with a rope-worked incline at Goathland to local iron mines. There was a passenger service, which was very similar to that on the Stockton & Darlington. The track was later converted for use by locomotives, and survives today as the North Yorkshire Moors Railway. The main centres of activity, however, were in the industrial north, together with proposals to link the L&MR to Birmingham and from there on to London. A new company was set up to run alongside the earlier Newcastle manufacturing company. Named George Stephenson & Son, it was intended to deal with all aspects of railway construction, from preparing for the Parliamentary Bill to completion of the line. Every promoter now seemed to want to use the new company, and work had to be shared out between George and Robert.

The first important new route was from Birmingham to the Warrington & Newton line, which in turn linked in to the L&MR. It was given a suitably imposing name, the Grand Junction. There was an initial survey by Robert that took the route through Sandbach, but that was opposed by the powerful Marquess of Stafford, so a second survey was made by Joseph Locke that passed by a hamlet that was destined to become famous in railway history, Crewe. The company was delighted by Locke's work on the line, but there was never any doubt that George Stephenson would be first choice for chief engineer. There was, however, a problem. Locke had been called in to report on the misaligned tunnels at Liverpool and inevitably had to be critical. Although he had been a good friend of Robert, he was now persona non grata to the Stephensons. A compromise was reached. George would be in overall charge of the whole operation and directly responsible for the southern section, but Locke would be resident engineer for the northern section of the line and would take over all responsibilities if George was absent. The next job was to appoint contractors.

Locke decided to divide the line into 10-mile sections and let each section separately. By 1834, everything had been agreed on favourable terms. Stephenson, however, was still bargaining with a loose body of contractors and, when the prices being paid on Locke's section were compared with those on the southern, there was a startling imbalance. One example was the estimated cost for a viaduct at Penkridge, which Stephenson had put at £26,000. When Locke was asked to provide an alternative, he came back with a figure of just £6,000. The company now had a dilemma. They had no wish to sack Stephenson, but it was clear that Locke was doing a better job. They suggested that they should become joint chief engineers. Stephenson was highly insulted at the idea he should share the honours with a man who had started off as his apprentice and promptly resigned, leaving the whole works in the hands of Locke. George still had other projects to keep him busy, including a line from Manchester to Leeds. By following the line of the Calder valley, the route taken by the earlier Rochdale Canal, major engineering works in crossing the Pennines were mainly avoided. There was, however, a tunnel that crossed the summit at Littleborough and, at over a mile and a half, was then the longest railway tunnel yet constructed.

With George's appointment to the Grand Junction, the job of chief engineer for the other major route, the London & Birmingham (L&BR), went to Robert. It is a tribute to his predecessor, Jessop, that the route chosen for the railway closely followed the one chosen for the Grand Junction Canal. When the Bill was passed, Robert moved to London, where he and the family remained for the rest of his life. At that end of the line, the railway had to make a deep cutting through the London clay at Camden Town. Houses had to be demolished and the whole scene was chaotic, as locals had to cope with a gash through the middle of their district that was almost unpassable until bridges were built. Moving north, the next major work was a mile-long tunnel at Watford. The construction was marked by a tragedy. According to an official report, a contractor had gone down in the night and removed the wooden supports for the roof so that the bricklayers could start work on the lining the next morning. While the night shift was at work, the tunnel caved in and eleven men lost their lives. The next obstacle to be faced came when the line reached the Chiltern hills, and once again Stephenson followed Jessop by digging a deep cutting using barrow runs.

Nothing caused Robert more trouble, however, than the massive Kilsby tunnel, which at its deepest was 132ft below the surface and 2,398 yards long. He had hoped it would be dug through solid clay, but it hit quicksand that had not shown up in the test borings. As fast as it was dug out, it flowed back in again. Robert took personal charge of the operations when the main

8.8 The barrow runs at Tring cutting on the London & Birmingham Railway. (From J.C. Bourne, *Drawings of the London and Birmingham Railway,* 1839)

contractor died. Two shafts were sunk and steam-powered pumps used to drain away the water at a rate of 1,800 gallons a minute. Even so, it was a long process, and the first 585 yards of tunnelling took eighteen months to complete. The rest of the work went more smoothly, taking just six months, and everything was completed by the summer of 1838. Robert also needed to build bridges across the Regent's and Grand Junction Canals. Constructing conventional bridges using wooden formers would have resulted in an unacceptable stoppage of boats on the canal. He looked at the possibility of using suspension bridges, but quickly dropped that idea, having seen how alarmingly one moved when a train crossed it on the extension of the Stockton & Darlington. He opted instead for iron girder bridges.

While still fully engaged on the London & Birmingham, Robert was consulted on another scheme, to link London to Brighton. Two rival schemes were put forward, but this time Robert lost to the opposition and was able to concentrate on other urgent matters. The site for the London terminus for the L&BR had not yet been agreed. The first option was for a site near the Regent's Canal and, when that was rejected, one near Marble Arch was proposed, which was also turned down, as was the third choice for a site at Maiden Lane. Eventually, agreement was reached for a station at Euston, close to the site where Trevithick had tried to persuade London investors to help finance his locomotive construction plans. The original station bore no resemblance to the present structure. The main train shed was simple, with pitched, glazed roofs carried on iron pillars, and just two platforms: one for arrivals, the other departures. The entrance in front of the station, however, was grand – a huge Doric arch. It was demolished on the orders of Minister of Transport Ernest Marples in 1961, in spite of widespread protests. The rail approach to the station was up a 1 in 68 gradient, which was considered too steep for the locomotives of the time and trains had to be hauled up by cable. The Birmingham terminus was at Curzon Street. This time the arch has survived, but the station was closed in 1966. A new station is currently under construction on the site.

The last decision to be taken for the L&BR was the provision of rolling stock. As Robert Stephenson was the chief engineer, it was considered inappropriate to allow him to also supply locomotives. The order went instead to Edward Bury, who was still in favour of four-wheeled engines. They proved to be hopelessly underpowered: it was recorded on one occasion that a train of forty-five goods wagons required an astonishing seven locomotives. The company had no option: they had to order Stephenson engines after all. The whole line was finally completed on 21 June 1838.

8.9 The original Euston station. (From *Osborne's London and Birmingham Railway Guide*, 1840)

As well as the work that was proceeding in building new lines in Britain, construction had already begun on rail construction in other countries. The French engineer Marc Seguin visited Newcastle in 1827 and ordered two engines for a line linking St Etienne to Lyons. A more unusual order came from the United States. The Delaware & Hudson Canal had to cross the Allegheny Mountains and the company built two inclines at each end of the summit level. When arriving at the foot of an incline, the boat was divided into two sections and hauled up to the top. There they were moved on a railed track and the company's engineer, Horatio Allen, visited Britain to see if this summit section could be worked by locomotives. He decided that was feasible and ordered four locomotives, one of which was supplied by the Stephenson works. A more conventional American railway, the Camden & Amboy, bought a Planet-class locomotive for its line, which was named *John Bull*. Not every potential client made the trip to Newcastle. Leopold I, the Belgian king, invited George and Robert to his country to give advice in 1835. There were long discussions on how a state railway system could be developed and the advice was so much appreciated that George was made a Knight of the Order of Leopold. Three engines had already been ordered the previous year. The two British engineers returned to Belgium for the opening of the line from Brussels to Ghent and this time it was Robert's turn to be knighted. Locomotives were also supplied to what was then Bavaria and France. The latter was a new design of Robert's. He was anxious to increase the size of the firebox to work with a larger boiler. He felt that increasing the diameter of the latter would put too much strain on the wheels, so he built a longer boiler and, to take the extra weight, all six wheels were placed under it, leaving the firebox overhung at the back. The long-boiler engines were an immediate success.

Most European lines were being built to the same gauge as the British lines, but an order came through from Czar Nicholas I, a railway enthusiast who as Duke Nicholas had visited the Middleton Colliery line. The gauge was described as 2 arshim 5 vershak. Aware that this might puzzle the British, they opted for the nearest equivalent – 6ft. One of the young men working at Newcastle was Daniel Gooch, who was very impressed by the engine, and we shall meet him again in the next chapter. The next order for wide engines came from the US for the 5ft 6in gauge New Orleans Railway. Unfortunately, America was going through a

8.10 The locomotive *John Bull* built in Newcastle for the Camden & Amboy Railway in America. The cowcatcher on the front was a later addition. (From *Dream City*, 1893: anon)

financial crisis and the order was cancelled. However, a new line was being constructed in the south of England under the direction of another young engineer, Isambard Kingdom Brunel. This was, of course, the 7ft gauge Great Western Railway. Brunel was having problems finding locomotives to work the line, so the two American locomotives were adapted and sent down to Bristol. Although they differed over the gauge question, Brunel and Robert became good friends, each supporting the other at difficult times in their careers.

In 1834 George Stephenson went to a meeting with the promoter George Hudson at Whitby. Hudson became known as the Railway King from his extended interests – it was only later in his career that he was exposed as a criminal misuser of funds. The two men had a mutual interest in developing railways in the north of England. Three lines were discussed – the Birmingham & Derby Junction, the North Midland from Derby to Leeds and the York & North Midland. All three were to be surveyed by Stephenson. It was during the construction of the Clay Cross tunnel on the North Midland that considerable mineral deposits were discovered. George Stephenson was one of a consortium set up to develop mines in the area and also limestone quarries at Crich. Extensive limekilns were built in the valley at Amber Gate. As part of the agreement, Stephenson agreed to build a narrow-gauge line from the quarries to the kilns and the Cromford Canal. It was at this time that he made his final house move to Chesterfield.

These schemes were all part of a plan of Hudson's to make York a major railway hub, at the heart of a route up the east coast to Scotland. At first, York had a modest station just outside the ancient city walls. Later this was replaced by a more extensive station and an arch had to be knocked though the city walls for the lines to reach it. It is doubtful if this would be allowed today, but Hudson was such a powerful figure at this time in local politics that no one opposed him. Robert was engaged in the north on a mineral railway, the Stanhope & Tyne, which was intended to serve extensive lime kilns at Stanhope. There was one major obstacle: Hownes Gill, a ravine 800ft deep and 1,140ft wide lay right across the line. Rather than go to the huge expense of building a viaduct, Robert devised a complex scheme of inclines at both sides of the Gill, worked by stationary engines. A far more important scheme came Robert's way when Hudson bought out all the shares in the North of England Railway, which would ultimately run from York through Newcastle to the Scottish border at Berwick. He was invited to survey the route.

It was at this time that tragedy entered Robert's life. Fanny had been in poor health for some time, but she was then diagnosed with cancer and died on 4 October 1842. They had never had children and now Robert sold the house they had lived in and moved closer to his London office to devote all his time to work.

In 1837 George Stephenson announced in a letter that he intended giving up business 'in the course of two or three years'. He was, however, still kept busy developing the Clay Cross Company, and was constantly being asked to advise companies planning new railways. One of these was the Preston & Wyre Joint Railway. The developers had plans not just for a railway, but also for developing a harbour at Fleetwood, and Stephenson approved the plans and suggested that they should also consider a coastal route from Liverpool to Preston that could later be extended via Maryport

8.11 The original York station.

to Carlisle. This was, he thought, the only practicable line, as going across Shap Fell would, he declared, be 'out of the question'. He was right in thinking the Preston & Wyre line would help in turning Fleetwood into a major port, although not as he predicted for passenger ships, but for a huge fishing fleet. He was wrong about Shap Fell, which was crossed later on the west coast main line to Scotland.

George did eventually ease off and he took more interest in his estate and gardens at Tapton House in Chesterfield. He devised a system for growing straight cucumbers by letting them grow into glass tubes, and he was an enthusiastic beekeeper. Railway matters were, however, never far from his mind, and one of the great questions of the day was: which was the better system, the Stephenson or the Brunel broad gauge. We shall look at that in more detail in the next chapter. Among the many schemes in which George was involved were two that took him across the Channel. The first was to Belgium again, this time to advise on the route for a line from the River Sambre near Charleroi to Vieux on the Meuse. He was honoured with a fine banquet at which the centrepiece was a model of *Rocket*. It was at this time that his second wife, Elizabeth, died in 1845 after twenty-five years of very happy married life. Soon after her death, he heard from an old friend, Sir Joshua Walmsley, that he was having difficulties with the Spanish government over a railway he was promoting to link Bilbao to Madrid and then onto the Bay of Biscay. George at once agreed to join Walmsley on a trip to Spain to look over the probable route. The journey through mountain passes was exhilarating, travelling in an open carriage, but their stay in Madrid was less successful. Tired of bureaucratic delays, they left the city with nothing settled. Two incidents on the return journey through France say a lot about George Stephenson. They came to a bridge across the Dordogne, and the engineer eyed it for some time before declaring that it seemed likely to collapse. He felt it his duty to go and see the local mayor and warn him of the danger. That gentleman listened politely but did nothing: a few months later, when a regiment of soldiers were marching across the bridge, it gave way, and many were killed. Later the two men came on a railway construction site and George was unimpressed by what he saw – 'their posture is all wrong'. He jumped out of the carriage and filled a barrow in half the time the navvies were taking.

Back home he was a keen advocate of good education, specifically in practical engineering. He supported mechanics' institutes and regularly spoke at meetings of the British Association for the Advancement of Science. When the Mechanical Engineering Institute was inaugurated in October 1846, Stephenson was elected president. He was more than a mere figurehead and regularly gave papers on a variety of topics, especially rail safety. After the death of his wife, Ellen Gregory, the daughter of a Bakewell farmer, moved in as housekeeper. In February 1848 they were married in a very quiet wedding, which even Robert was not told about. It was to be a short-lived arrangement. In the July of that year he became seriously ill with pleurisy and he died on 12 August. He had always hated ostentation and pomp, so when he was buried in the graveyard at Holy Trinity, Chesterfield, the gravestone was quite plain, just inscribed with his initials.

During his father's last years, Robert was as busy as ever. In 1838, George had been asked by the Chester and Crewe Company to survey a possible line from Chester to serve the mail services to Ireland. Two possibilities were advanced, one to what was then a small fishing village on the Llyn Peninsula, the other to the port already used for the Irish Mail, Holyhead.

They opted for the latter and Robert was asked to survey the route the following year. The first part of the line from Chester only presented problems of the type already overcome elsewhere: a 405-yard tunnel at Chester, a viaduct and a bridge over the Dee. After that it was a circuitous route that hugged the coast as far as Great Orme Head and then turned towards Conwy. There it would have to cross the broad estuary of the Dee, but beyond that was the far greater obstacle of the Menai Straits. Telford had already bridged the latter gap, but Robert could not contemplate using the same solution by building another suspension bridge.

Robert's first idea was to make use of the existing bridge at Menai. One roadway would be laid with rails. Trains arriving from Chester would be uncoupled, and the trucks or carriages would be pulled across in batches by horses, and then recoupled to a second engine on the opposite bank to complete the journey. There was, however, a problem. The approach to the bridge was too steep for locomotives, so a long, gentle embankment would be needed. The bridge owners were happy to lease out part of the roadway, but insisted that any embankment had to be temporary. That made the whole scheme impractical. Robert next considered making a bridge of two cast iron spans, each 350ft across, but there was no obvious way of constructing it. Brunel came in with a suggestion by which the parts could be built out from either side of a central pier so that they would always be in balance, but that idea was rejected. Robert's next idea was to construct the arches on floating pontoons, so that they could be jacked into place when completed. The Admiralty immediately vetoed that idea as the pontoons would interfere with navigation. Suspension bridges were again considered, and Robert began to consider ways in which they could be made structurally sound. He returned to an idea that he had first used for a bridge over the River Lee at Ware. Here he had used wrought iron boiler plates to create cells that were fastened together to make a very rigid structure. He saw these, at first, as being still part of a suspension bridge, He put the idea forward and there was some criticism, with some engineers suggesting that an oval-shaped tube would be more efficient and stronger. But Robert realised that in fact the tube, with its thick cellular metal plates at top and bottom of his rectangular tube, with the thinner plates at the side was, in effect, a very large box girder. His new idea was to have a large enough tube for the trains to run inside it, and it should not need suspension chains at all.

This was an untested idea so Robert sought the advice of William Laird, who was building iron-hulled ships at Birkenhead. Laird was able to reassure Robert and give a practical example in support. An iron ship, the *Prince of Wales*, was being prepared for launch when a cleat gave way. The vessel plunged into the river, the bows resting on the riverbed and the stern stuck on the quay. However, the iron frame remained intact. This was just the encouragement Robert needed and he began to draw up plans for tubular bridges at the two crossings. The larger tubes for Menai were to be 450ft long, 30ft high and 18ft wide, and weighing 600 tons each. Before work could begin, tests were made using ⅛th-scale models, starting in December 1845. Only when these proved satisfactory could work get under way.

It was decided to start with the less problematic Conwy bridge. There were to be two tubes to take a single track each. The wrought iron plates were shipped in from Liverpool and riveted together on site. Once completed they were to be loaded onto pontoons, floated out into the river and jacked into place. It was an anxious time for Robert when the day to move the first tube arrived, and Brunel came along to lend moral support. Things did not immediately go well when one of the pontoons drifted out of line and it could not be set back in the right position until high tide on the following day. That proved difficult, but eventually everything was secured, the tube was raised and, on 18 April 1848, Robert rode on the footplate of the first train to pass through the tube. Single-track services were soon started and work went ahead to position the second tube. Work now could begin on the more dangerous job of constructing the Menai bridge over the fast-flowing tides of the Straits.

The same technique of constructing the tubes on site was used. Work on the Conwy had exhausted Robert and he took a short break at Windermere to recover, but was soon back in Wales. Once again there were setbacks in moving and lifting the giant tubes. The movement of the giant tube was controlled through capstans on the shore, and the first requirement was to secure one end of the tube on its pontoons to the Anglesey pier. The foreman in charge, Rolfe, realised that the tube was swinging too wide and was going to miss the pier altogether. He hastily tried to correct the swing with one of the capstans, but the force of the tube was too great and the machine was dragged from the ground. He managed to grab the cable, shouted for help and men ran to his

FLOATING THE
HOLYHEAD and CHESTER RAILWAY TUBULAR BRIDGE,
AT CONWAY.
6th March 1848.

8.12 Raising the tube for the railway bridge at Conwy. (Elton Collection, Ironbridge Gorge Museum)

assistance and managed to save the situation. After that, things went more smoothly and the whole bridge was opened on 5 March 1850.

While work was still going on at the Conwy bridge, there was a disaster on another part of the line. Robert had used a similar bridge to that over the Lee, using both wrought iron and cast iron to create a girder bridge across the Dee near Chester. It had passed inspection, but on 24 May 1847 the driver of a train heard a cracking noise as he crossed the bridge. He at once opened up the regulator and the locomotive was clear when the bridge collapsed, but the coupling broke and the tender, van and carriages plunged into the river. Altogether six people died and sixteen were injured. Inevitably, there was an inquiry, and several eminent engineers gave evidence to say that they considered the bridge was safe and there must have been some other cause. The inquest jury brought in a verdict of accidental death with no blame attached to any individual. Robert was extremely fortunate, for there is no doubt that the combination he had used of the two types of iron was fundamentally flawed. Had he been found culpable, his career would have been at an end.

Robert's other main project in these years was the line to Berwick, and the first decision he had to make was how to cross the Tyne to Newcastle, standing high above the river. He opted for a double-level high bridge, consisting of five stone piers linked by bow string iron decks, the lower for use by road vehicles, the upper for the railway. This system used cast iron arches for the 'bow' and wrought iron 'strings' connecting them to the decks. The foundations consisted of timber piles encased in concrete. The newly invented steam hammer was used to drive in the piles, and the bridge was completed in 1849. The route north from Newcastle was comparatively straightforward. Even the crossing of the Tweed at Berwick offered no new challenges as the river was not navigable, allowing a conventional stone viaduct to be built. It was, however, on an imposing scale, consisting of twenty-eight stone arches lined with brick, carrying the lines 126ft above the river. Pile driving proved difficult as they had to be driven through 40ft of gravel before firm ground was reached. Impressive statistics were given: 2½ million bricks were used and 8 million cubic feet of stone. The official opening in August 1847 was attended by Queen Victoria and Prince Albert, and it was named the Royal Border Bridge, though strictly speaking the actual border is a few miles to the north. With its opening, there was now a through route from London to Edinburgh.

By the end of the 1840s, Robert was increasingly involved in overseas railway projects. In 1847 he produced plans for Norway's first steam railway that was to run from Oslo for 40 miles to Eidsvoll. Although the plans were approved, there was a cash shortage and work only got started in 1850. It was a modest single track that carried more timber than passengers. Robert was also approached by the Swiss, who had started the process with the formation of the Swiss Northern Railway Company in 1836, which by 1848 had constructed a very unimpressive 14½ miles of track. At this point the government invited Robert to put forward proposals for how a modern rail system could be provided to serve the whole country. He suggested that they needed to think in terms of international connections as well as local, with lines from Basle to the French border and to join the German network near Lake Constance – a third proposal to connect with Italy would have involved crossing the Alps to Lugarno, which he considered was beyond the technology of the day. The other routes all

8.13 The inaugural run of Norway's first steam railway. (Author's collection)

lay along river valleys. Robert sent one of his staff, Henry Swinburne, to carry out the appropriate surveys. Having been to Norway, Robert was now invited to Denmark. He went with the railway contractor Morton Peto to Copenhagen to discuss the plan with the Danish government. There were two routes agreed, one from Copenhagen to Roskilde in Zealand and the other from the coast of Schleswig Holstein to the Baltic. Robert's involvement was limited to the planning stage. He completed the Scandinavian trilogy by advising on plans for Sweden, where again most of the work was left to another engineer, William Lloyd, who had joined the Stephenson company in 1848. He surveyed some 700 miles of line, the first section of which opened in 1856.

Now Robert was to travel even further afield, leaving Europe to look at railway possibilities in Egypt. India was now in practice largely under British control in the form of the powerful East India Company. Sea routes, before the opening of the Suez Canal, involved the long voyage around the tip of Africa. The P&O shipping line shortened the journey time by unloading passengers and goods in Egypt and moving everything overland to the Red Sea, where another vessel would complete the journey. However, the advantage of a rail connection was obvious. In 1848 Robert visited Egypt to discuss ideas, but nothing came of it. He was back again in 1850 and this time there was a positive proposal for a line from Alexandria to Cairo. Robert's preferred route was straight down the Nile valley, but the Egyptians wanted the route to go via Tanta in the Nile delta. This involved crossing several waterways, one of which, the Karrineen Canal, was crossed by a tubular bridge similar to those used on the Holyhead line. Elsewhere, bridges with central, swinging sections were used. The Nile itself presented a greater problem, and for many years everything was taken across by a conventional ferry. Later a chain ferry was introduced, with a railed deck to carry trains. A huge labour force of around 24,000 was employed and work went on rapidly. Robert had hoped to be given the job of the extension of the line from Alexandria to Suez but the contract went instead to M. Mouchelet. When Robert later came to see the works, he was notably unimpressed – 'a huge engineering blunder'.

India itself was slow to join the railway world. Inevitably, when plans were first considered, Robert was consulted. He gave his first report in 1847, but the finances never appeared. The plans were revived in the 1850s, and the first modest line of just 30 miles from Bombay (Mumbai) to Kalyan was approved, with the first section as far as Thala opening in 1853. The real objective was not a coastal route, but one that would link Bombay to the Indian interior. However, that would involve climbing the Western Ghats, which rose 2,500ft above the coastal plain. Stephenson suggested employing James Berkley to go to India to see what could be done. His proposal was dramatic, involving severe gradients and a line that would zigzag its way up the cliffs, with many reversals. It received Robert's enthusiast backing: 'I should be proud of being the author of the plans.' Building the line was incredibly difficult and conditions atrocious: of the 40,000 Indian workers, around a third died from disease.

One other major overseas project involved Robert. Alexander McKenzie Ross, who had worked with him at Conwy, had been asked to come up with plans for bridging the St Lawrence River at Montreal, which would require a span of over 6,500ft. He proposed a tubular bridge and took his plans to Britain for Robert's approval. Robert not only thought the plans sound in principle, but went across to see the site for himself before giving formal approval. It was an immense project, with the tubular bridge to be built in twenty-five sections resting on twenty piers, standing 60ft above the high-water mark. The approaches to the bridge were on embankments faced with large stones. The actual iron work was manufactured at Birkenhead under the supervision of Robert's cousin, George Robert Stephenson. Working conditions on the site at Montreal were often atrocious, with winter temperatures dropping as low as -30°C, and the piers had to withstand a battering in spring when the thaw came and vast chunks of ice were swept down the river. The great bridge was finally opened for traffic in 1859.

Robert Stephenson was always in great demand as railway promoters from around the world sought his advice. He managed to find escape when he purchased a handsome yacht, *Titania*, built at John Scott Russell's shipyard on the Thames. Once aboard no one could bother him – he called it 'the house that has no knocker' – and found it ideal to use as his temporary home when he spent the winter of 1850–51 in Alexandria. When in Britain, he continued to advise on various railway plans and, rather surprisingly, was also consulted on water supply for several cities, including London and Liverpool. Even more surprising was an invitation that came from George Hudson to stand as Conservative candidate for Whitby, an offer that he accepted. This was such a safe seat that the Liberals did not even bother to put up a candidate. Why he accepted we shall never know, and he seldom visited the constituency. Apart from being Conservative, he was also very conservative in his attitudes. He spoke out, for example, against a Liberal proposal for educating the working classes, arguing that artisans only needed to be taught what they needed for their trade and nothing more.

One piece of Parliamentary business in which he took an active interest came about through his membership of the Society for the Encouragement of Arts, Manufacture and Commerce. The French had staged a splendid exhibition of arts and manufacture in a specially built hall in the Champs-Élysée in Paris. It was decided that London should have a similar exhibition. A committee was set up by the society to consider the matter, to which Robert was elected. Nothing much came of it, but Prince Albert became an enthusiast and, thanks to his involvement, it became an official government scheme. Robert was now one of the MPs appointed to the special commission charged with choosing a design for the exhibition hall. Among those whose proposals failed to win approval was one from Stephenson's friend, Brunel. Edward Stanley Ellis was the chairman of the Midland Railway and an MP and, one day, when having lunch with

Joseph Paxton, well known for the great glass house he had designed for the gardens at Chatsworth, he mentioned the trouble they were having finding a good design. Paxton sketched a rough outline of a building consisting of panes of glass in a complex wrought iron frame. Ellis took the design idea to Robert, who supported it enthusiastically. The final result was the famous Crystal Palace that housed the Great Exhibition of 1851.

In the autumn of 1857, Robert took a long break from work, sailing around the British coast in *Titania* with friends. At the end of the year, Brunel was preparing for the launch of his immense vessel the *Great Eastern* into the Thames. Robert went along to offer his support, but slipped and found himself up to his waist in muddy water. In spite of his soaking wet clothes, he insisted on staying with his old friend. Things did not go well, and the ship was only launched at the end of January 1858. At this time, both men were suffering from bad health and exhaustion from an unrelenting work schedule. Robert decided to escape the following winter by sailing once again to Egypt. Brunel joined him later, and they had Christmas lunch together in Cairo. Back in England, he was invited to Norway for the opening of the line from Oslo in August 1859. It was a grand occasion and Robert was awarded the Order of St Olaf. He was clearly unwell, and the party hastily laid plans to return to England. They eventually landed at Lowestoft, where Robert received the sad news that Brunel had died. He was soon to follow him, dying peacefully at home on 12 October 1859.

George and Robert Stephenson dominated the early years of railway development. Although George was not a great innovator, he kept faith in the whole concept of steam railways when others seemed to have lost interest, and was responsible for constructing lines that changed the system from one that did little more than move coal from ports to rivers and canals to one that linked cities and brought a new form of passenger transport to the world. Robert did as much as anyone to transform the locomotive: in designing *Rocket* he introduced all the elements that were needed for future development. They were true pioneers.

9

ISAMBARD KINGDOM BRUNEL

Brunel is the one engineer that almost everyone in Britain has heard of, and probably regarded as the greatest, even though he was guilty of far more serious errors of judgement than others whose achievements have been described here. His father, Marc Brunel, was French and served for a time in the navy, but was forced to go into hiding during the French Revolution as he had openly expressed royalist sympathies. During that time he met a young English woman, Sophie Kingdom, who had been sent to France to learn the language. She had also been caught up in the revolution and was sheltering in the same house. They fell in love, but Brunel was forced to flee to America, leaving her behind. From America, where among other activities he had surveyed canals, he came to England and at once went to look for Sophie. They were married in 1799.

Marc had a remarkably successful career, devising one of the first factories to use mass production techniques. It was the block mill at

Image: Isambard Kingdom Brunel. (Science & Society)

Portsmouth, where a whole range of machines designed by him and made by Henry Maudsley were used to manufacture the blocks that were a vital part of a ship's rigging. He also developed new, improved machinery for sawing timber. The Brunels had two daughters, then, on 9 April 1806, they had a son, Isambard Kingdom. To his father's delight, the young Isambard showed an early interest in all things mechanical. Marc taught him drawing, a skill that he felt was essential in any engineering career, and after day school the boy was sent to Dr Morrell's Academy at Hove as a boarder. In a letter home to his mother, he mentioned that although he enjoyed reading Horace he preferred Virgil, and told her he had been drawing in the town and making maps. He was clearly a bright scholar.

Marc Brunel was not impressed by further education in Britain, which as Robert Stephenson had discovered was generally of little help for anyone thinking of a career in engineering. So, Isambard was sent to France and enrolled at the Henri Quatre Lycée in Paris. From there he moved on to take an apprenticeship with Louis Breguet, one of the most accomplished makers of scientific instruments and timepieces in France. There he learned a great deal about precision engineering and thoroughly enjoyed the experience. At 18 he had hoped to go to the Polytechnic in France, but it was only open to French students and, in spite of having a French father, he was not admitted. He returned instead to England to begin working with his father, who in the meantime had been kept busy on a number of major engineering projects.

The most important of these was the revival of the plan for a tunnel under the Thames. The new tunnel was to run from Wapping to Rotherhithe and was approved by a Parliamentary Act of 1824, with Marc in charge. The first task was to carry out a series of test borings to ensure it was going through solid ground, and all seemed satisfactory. The plan called for shafts at either end with an internal spiral ramp that could be used by both carriages and pedestrians, joined by a double-arched tunnel to allow for two-way traffic. Marc had designed a new device, the tunnel shield, to enable tunnelling to be carried out efficiently and safely. This was a metal structure, with thirty-six compartments, each 7ft high and 3ft wide, that would enable thirty-six men to work at excavating the face. Each compartment had a set of heavy oak boards that were held in place. The miner would start by removing one board, then excavate the ground in front of him to a depth of 4½in. Then he would replace that board and remove the next. When the whole face had been dug away, the shield was moved forward. At the back of the shield was a platform for bricklayers to line the tunnel, and spoil was removed from behind the shield to the shaft. The tunnel shield has been the basis for major tunnelling schemes ever since.

Once the shaft had been completed, work could begin on the tunnel itself. The test borings suggested that all should go well but, as Trevithick had discovered earlier, although much of the work was through solid clay, there were always patches of loose soil and quicksand. Leakages from the river came in on a regular basis, often leaving the men knee deep in water. At this time, the Thames was seriously polluted as all the sewage from the city simply discharged into the river. The theory was that it would be washed away on the ebb tide: unfortunately, it was just as liable to wash back on the flood. As a result, conditions in the tunnel were so bad that many men became ill, including the resident engineer Armstrong, who resigned. Isambard was now promoted to the post.

In 1826, Richard Beamish joined the team to supervise the work, and almost at once had to deal with a major problem when silt started to flood into the workings along with the water. He and Isambard were kept busy for five days and nights, grabbing what sleep they could, to ensure that the flow of silt was plugged and only water was coming in that could be pumped clear. Both had to take time off to recover. Worse was to follow. At 2 a.m. on 18 May 1827, Beamish was taking over from Isambard when he became aware that, once again, there seemed to be danger of another serious flood. At the time, the company had allowed visitors to see the tunnel, and a special shaft had been opened up for them. Beamish had to show an aristocratic lady the scene and, as soon as she was gone, he changed into waterproofs and hurried back down. Almost at once, the river burst through. Beamish rushed to get the men out and he met Isambard at the foot of the shaft and looked back at a scene of devastation. Once at the surface, it was discovered that one workman was still missing. Isambard grabbed a rope, slid down an iron stay and helped haul the man up to safety. A roll call was taken and everyone was now accounted for.

Investigation with a diving bell revealed what had happened. They had reached a point below the river where, for centuries, ships had dropped anchor, scooping out a vast hole in the riverbed. They now had the task of filling the hole, a job that took them until November. Once the tunnel had been pumped dry, it was discovered that no real damage had been done and work could continue. As a mark of confidence that all was now well, a great banquet was held in the tunnel, organised by Isambard, as his father was not well. It proved to be a premature celebration. On 12 January, there was another flood. This time Isambard was in the tunnel. He was temporarily trapped when a timber fell on his leg, but he managed to extricate himself and to reach the foot of the main shaft, only to find it was blocked. He made his way to the visitors' staircase, where he was caught by a great surge of water that carried him up and out to the surface. He was doubly fortunate, for Beamish, realising what was happening, had opened the door at the top that was normally kept locked. Had he not done so, Isambard would have been swept back down and drowned. Others were less fortunate: six workmen died in the flood. Efforts were made to repair the damage and plug up the hole, but the money had run out and work was abandoned.

Both the Brunels were now unemployed and looking for new projects. Two requests came their way for bridges. One was to cross the Vistula in Warsaw, the other for the Avon Gorge at Bristol. Marc opted to take on the prestigious structure in the heart of a capital city, leaving Isambard to take care of the Bristol project. This had its origins with local wine merchant William Vick, who had left a bequest of £1,000 with instructions that it should be invested, and when the interest had brought the capital to £10,000, the money should be used to construct the bridge. He had died in 1754, so it was a long wait before the sum reached £8,000 and it was decided work could be contemplated. A competition was announced to find the best design. Marc had already had experience of building suspension bridges, and this was the idea taken up by his son. The entries were to be judged by Thomas Telford. There was a constant interchange of letters between Marc and Isambard, which resulted in three designs for suspension bridges with spans ranging from 760ft to 1,160ft. But Telford had serious concerns about the safety of any suspension bridge over 600ft long, following his own difficulties in stabilising the Menai bridge. None of the Brunel designs qualified, and all were rejected – as were those of all the other entrants.

The bridge promoters now asked Telford to design a bridge himself, which he agreed to do. The result was an extraordinary structure. There were two towers that rose from the river bank to the top of the gorge, designed in an elaborate Gothic style, with the suspension bridge slung between them. It was widely ridiculed and rejected. With Telford no longer around to demand limitations, the contract went to Isambard in March 1831. He was, however, forced to compromise, reducing the span to 630ft by building an abutment out from the Leigh Wood side of the gorge. The two towers were to be built in the Egyptian style. The chains were long links, joined by a single pin: lighter than those used at Menai.

Work got under way with an official ceremony in June of that year, but it was obvious that more funds would be needed to complete the work. It was proposed to make an appeal for finance in October, but the timing could hardly have been worse. The House of Lords had thrown out the Reform Bill that would have extended the vote to more of the population and removed rotten boroughs such as Old Sarum, which returned a member of Parliament despite the fact that it now consisted of just one farm. Disturbances in Bristol led to full-scale riots and special constables were appointed, one of whom was Isambard. They were quite unable to control the situation and parts of the city were destroyed before the dragoons were called in and brutally settled the matter: six citizens were killed and eighty-six injured. Any available money was now needed for repairing the riot damage, and the bridge project languished, with nothing to show for the efforts apart from one tower on the Clifton side. Isambard was, however, soon to find more work in Bristol.

The floating harbour, designed by Jessop, was a success, but there was a major problem with silting. Isambard suggested increasing the size of the dam on the Avon to divert more water into the Cumberland Basin, building a stop gate at Prince's Street and using a drag boat to remove the silt. The works were put in hand under Isambard's direction and he designed an appropriate drag boat. A drag boat to this design, *Bertha*, was completed in 1844. The vessel had a metal plate attached that reached down to the bottom of the harbour. It used a single-cylinder steam engine on board to pull the vessel along by means of a chain stretched across the harbour. The metal plate scraped up the silt and took it to the quay, where it could be scooped out. The vessel was used at the docks in Bridgewater. During Isambard's time working on the harbour scheme he became friends with the Quay Warden, Captain Christopher Claxton, who was to become a highly valued associate.

Work, eventually restarted on the Thames tunnel, with Marc again in charge, but Isambard no longer took any part in the work. There was now a new plan being prepared in 1832 in Bristol for a railway to London. As with other lines, the first essential thing to do before anyone could apply to Parliament was to have a survey made. Two names were initially put forward: Henry Price, whose proposal did not appeal, and William Brunton, designer of the failed 'walking engine'. Neither inspired confidence. One of the Dock Committee, Nicholas Roch, suggested Isambard should be considered for a part in the process. Eventually, it was agreed that a local land agent who had been involved in planning the Bristol & Gloucester Railway should take on the task with Isambard. It became clear that Isambard was by far the more competent of the two and was soon taking the leading role.

The first part of the route was obvious, following the line of the Avon to Bath. The next big question was whether to go south or north of the Marlborough Downs. Isambard looked at both routes and decided that the northern was far better, passing through Swindon to Reading and the Thames valley. He put his plans to the two committees, one in Bristol, the other in Bath. Once these had been accepted, he realised that he would no longer be able to share an office with his father and would need a place of his own. He moved to 53 Parliament Street, Whitehall. When making the initial surveys, he had frequently been frustrated by not being able to get transport when he needed it and in finding suitable inns for the night. What he needed, he decided, was a britzka, a long four-wheeled vehicle, with seats that could be folded down at night to make a bed – the nineteenth-century equivalent of the camper van. He had one made and painted black, earning it the nickname 'the flying hearse'.

The time came to present the Bill to Parliament for what was no longer mundanely called the Bristol & London Railway, but had been named the Great Western Railway. There was a problem, however, in that the developers had not yet got enough investment to cover half the estimated costs of completing the whole route, which was what Parliament required. They came up with an unhappy compromise, suggesting that they got an Act for building a bit of it in the east and a bit more in the west, and they would come back later for another Act to fill in the middle. Not surprisingly, the idea was rejected, so there was a pause until enough funds were available. There was the usual opposition from affected parties, notably the Kennet & Avon Canal. Then other schemes were put forward, including one that would follow the southern route that Brunel had rejected. One of the odder objections was that the proposed tunnel at Box was impossibly dangerous. A professor of astronomy, Dr Dionysus Lardner produced figures to show that because the line in the tunnel was on a gradient of 1 in 100, if the brakes failed when going down it the train would emerge at a speed of 120mph, which he claimed no human being could survive. Brunel calmly pointed out that he had made no allowance for friction and air resistance, and when they were taken into account the speed was nearly halved. In the event, all the opposition failed and the Act for the GWR was approved in August 1835.

Isambard did enjoy an active social life in spite of the pressures of work. He was introduced to the Horseley family, who lived in Kensington, close to the present Notting Hill Gate tube station. They were a very talented family. The father, William, was a musician and composer, one of whose works is still well known, his setting of the hymn 'There is a Green Hill Far Away', and he was a founding member of the Philharmonic Society of London. There were two sons, one who became a Royal Academician and the other who followed his father's career as a musician. Of the three daughters, Fanny died young, Sophie was a talented pianist and the eldest seems not to have shared in the handing out of talents. The family thought her rather haughty and nicknamed her 'the Duchess of Kensington', but she was the family beauty. Many famous musicians came to visit and one of them, Felix Mendelssohn, seems to have become very attracted by Mary, but nothing came of it.

Isambard also found Mary attractive, but he was determined not to get married until he was in a position to offer his wife a good, secure life. With the passing of the GWR Act his future was looking bright, and when he was also appointed as engineer for extensions from Swindon that would take a line up to Cheltenham with a branch to Cirencester, he felt ready to propose to Mary. He moved to a new house at 17 Duke Street, near St James's Park in London, and they were married in July 1836. After two weeks' honeymoon in Snowdonia, Isambard was back to work.

Isambard now began to plan the track on which the trains would run. He had taken a trip on the Liverpool & Manchester and had been unimpressed. It was far from smooth, so he decided simply to ignore what Stephenson and others were doing and to try and find the best possible system to ensure the most comfortable ride. He changed everything. For a start, instead of the usual edge rails, he decided to use bridge rails, shaped in cross section like a wide-brimmed hat, with a tall, rounded crown. To ensure a rigid running service, he had the rails mounted on longitudinal wooden sleepers, running the full length of the rail. These were joined by cross ties that were spiked into place, and the whole structure ballasted. The most radical decision, however, was to change the gauge. He thought 4ft 8½in far too narrow for comfort, and decided to build his line to a 7ft gauge, although it finished up as 7ft ¼in. It was certainly rigid, too much so in fact, with very little give in it.

At the same time as he was planning the GWR route, work restarted on the Clifton bridge. A 1,000ft-long wrought iron rod was hauled across the gorge, under which a simple carriage was suspended that could be hauled across by hawsers. When Isambard made the first trip across, the bar sagged, the roller jammed and he simply clambered up the supporting wires to free it. The first job was to construct the tower on the Leigh Wood side, a laborious task that was only completed in 1840. Three years later the funds ran out and once again everything came to a halt. Clifton bridge has always been regarded as one of Isambard's great works, but in fact it was only completed in 1864, long after his death, by the engineers William Barlow and Sir John Hawkshaw. It is substantially different from the original Brunel design: the towers have been altered, the deck is more substantial and there are more suspension chains used. We cannot know how the original would have performed if it had been built as planned. There is always the possibility that Telford's caution might have proved right, although that is unlikely.

Back at the GWR planning stage, decisions had to be taken on where to site the two termini. At the London end, a temporary timber station was built on what was then the outskirts of the city at Bishops Bridge Road, Paddington. At Bristol, the city authorities, unlike those of York, refused to allow the lines to pass through the old city walls, so a site was chosen at Temple Meads. Isambard chose a mock medieval style for the station, with a train shed that boasted a mock hammer-beam roof. There were two platforms, one for arrivals, the other for departures, and passengers could cross between the two via the undercroft. The company offices were also part of the structure. The original station is still there, but became redundant when a new through station was built to allow the GWR to extend eastward to Exeter.

9.1 The first Bristol station at Temple Meads. (From J.C. Bourne, *History and Description of the Great Western Railway*, 1846)

The first major obstacle to be faced was the tunnel at Box. The technology that was needed was the same as that already described in the previous chapter, but this was on a far larger scale, simply because the tunnel had to be wide enough to take two broad-gauge tracks. The bore was 25ft in diameter and was 1⅓ miles long. The navvies had to blast their way through oolitic limestone, and also remove sections of fuller's earth. When the work was completed in 1841, statistics were published that make impressive reading: 414,000 cubic yards of spoil were removed; 54,000 cubic yards of brick and masonry were used; each week a ton of explosive for blasting was expended, along with a ton of candles to light the work. The workforce varied in number, but there were never fewer than 1,000 employed. What the official statistics do not reveal is the number of men who died during construction, but it has been estimated that the figure could be as high as 100. One story that has been accepted for years is that at dawn on Brunel's birthday, the light shines straight through the tunnel. Rather sadly, recent research has shown that is not actually true.

After Box, the line then took a great sweep around the north of the Marlborough Downs. It passed through the village of Swindon, a spot

9.2 A Firefly-class locomotive leaving Box tunnel, being given the all clear by the signalman. (From J.C. Bourne, *History and Description of the Great Western Railway*, 1846)

selected as the site for the works, where locomotives and rolling stock could be built and repaired. To accommodate the workforce, a whole new town was built and with housing that was of a very high standard for the time. The new town still remains and, although there have been improvements to bring it up to modern standards, it is still an attractive, well-planned area. From here the line continued its great sweeping curve to join the Thames valley between Didcot and Reading.

The next challenge was the crossing of the Thames at Maidenhead. The river authorities insisted that there could be no more than two spans, so each would have to be 128ft across. Brick and masonry bridges at the time were virtually all built with semi-circular arches, so the crown would have had to stand so high above the river that Brunel would have had to use embankments at either end to bring the tracks up to bridge level. However, he knew that bridges did not have to be built that way. While working on the Thames tunnel, his father had constructed a section of an experimental bridge at Rotherhithe, which was 100ft long, but the crown was only 10ft 6in high. Marc sent his son notes and suggestions, and the new bridge was designed with a central pier on an island and with the required two arches, which stood just 10ft 6in above the water. If semi-circular arches had been used, the crown would have been 64ft high. Many experts were convinced the structure would collapse, and they seemed to have been proved right when some of the centring was removed and the brickwork began to crumble. It turned out, however, that the mortar had not been allowed to dry and the damage was easily made good. However, when both arches were completed, the centring was left in place. The sceptics declared that Brunel was too scared everything would collapse if he removed it. Then a storm washed away the centring and the bridge stood firm, as it remains to this day. A second viaduct had to be built across the Brent valley at Hanwell, but this was simply a conventional eight-arched structure.

Brunel's plans did receive one setback. He had intended to build his line through Eton, but the famous school objected. They seemed mainly concerned that their boys might jump on a train and go into London, where all kinds of temptations might lead them astray, so the route had to go through Slough instead. The final approach to London closely followed the line of the Paddington arm of the Grand Junction Canal. The first train to make the journey between Bristol and London ran on 30 June 1841, but by then the line had already been extended to the west and had reached Bridgewater.

If the completion of the railway was a triumph for Isambard the civil engineer, the same could not be said for his other role in working out plans for the locomotives to run on the line. He laid down strict conditions for engine design: the piston speed must not exceed 280ft/min and weights must be limited to 8 tons for a four-wheeled engine and 15½ tons for one with eight wheels Their working speed had to be at least 30mph. This was bizarre: Robert Stephenson already had successful engines with piston speeds of 500ft/min and his Planet-class engines were 10 tons on four wheels. The first order went to T.F. Morrison of the Hawthorn works in Newcastle. In order to compensate for the low piston speed, he used enormous drive wheels, but that made them exceed the weight level. To reduce the load he had the power unit on separate wheels from the boiler, with the two connected by a flexible steam hose. Two geared engines were also delivered from the Haigh factory in Wigan. None of them could be described as even remotely satisfactory, simply because the required performance was impossible given Brunel's set limits.

Fortunately for Brunel, a young man appeared on the scene who was to change everything. Daniel Gooch had proved his extraordinary abilities by being made locomotive superintendent at the Stephenson Forth Street works when just 20 years old. It was while he was there that he worked on an engine, built for export – there is some doubt as to whether for Russia or America – for a 5ft 6in gauge line. Gooch was very impressed and became an enthusiast for a broader gauge. This was, of course, anathema to the Stephenson camp, so Gooch applied to join Brunel. He remembered that there was a 5ft 6in gauge engine that had been built at Newcastle for America, but the order had been cancelled, and he realised it could be adapted for the Great Western. The engine was purchased and the necessary changes made. It was named *North Star* and was certainly a huge improvement over the earlier engines. It arrived just in time, for there was a faction of shareholders based in Liverpool who were unimpressed by Brunel and all his works. They saw the old engines stumbling along the broad gauge, while in Liverpool they were used to the far more efficient Stephenson locomotives and the service they provided.

A decision was taken to invite three engineers to report on the line and its engines. Two sent in reports that were quickly dismissed, but the third by Nicholas Wood was taken seriously. Wood decided that he needed scientific advice to help him in his task and the man he chose was Dionysus Lardner, who had made the very unscientific pronouncements on travel through Box tunnel. He now declared that it would be scientifically impossible for *North Star* to reach a speed of 30mph. In fact, the engine had not been performing as well as expected, but an investigation showed that the problem was a badly placed and inefficient blast pipe. Once that was rectified, the locomotive was given a trial and covered the journey between Maidenhead and London at an average speed of 38mph while hauling a 42-ton load. Dr Lardner was defeated and so too were the Liverpool faction. Brunel won a vote of confidence. Thanks largely to Gooch, the broad gauge was saved – and so too was Brunel's career. Gooch went on to improve on the first engine with a new locomotive called *Firefly*, and by 1846 the first engine steamed out of the Swindon works. Named *Iron Duke*, it was more powerful, with a 2-2-2 wheel arrangement, and achieved speeds of around 50mph on a level track. Engines of this class were to serve the broad gauge well.

Until the time the line was completed and a reliable engine provided, Brunel was subjected to regular criticism. At one meeting a member of the committee went so far as to suggest that the line from London to Bristol would never be completed. The engineer famously replied with typical bravado – why stop at Bristol, why not go on to New York? Most ignored the comment, but one member, Thomas Guppy, took it seriously. Brunel was never one to resist a challenge, especially one that everyone else thought ridiculous. A new company was formed, with Brunel, Guppy, Captain Claxton and another GWR visionary who was excited by the idea, Peter Maze. As a result of the group's first meeting, it was named the Great Western Steamship Company. Brunel, already working continuously on overseeing the railway, now had transatlantic travel to worry about as well.

As soon as the plans became widely known, they aroused equally strong feelings of enthusiasm on one side and scepticism on the other. A small steamer, the *Savannah*, had already a made a crossing from America, but only used steam for part of the voyage – when the wind was favourable, she worked under sail. The paddle wheels were designed so that they could be lifted out of the water and stowed on deck. The whole journey took 659 hours and steam was only used during eighty-five of them. Could a vessel steam all the way? The problem appeared to be that it would be difficult to make a vessel economically viable if too much space was taken up with coal for the boiler. As ever, Dr Dionysus Lardner was ready with his own brand of expertise. There was no point in doubling the size of the ship because you would simply double the amount of coal needed and there would be no benefit, he argued. The learned doctor had once again got his sums wrong. A ship's weight is supported by the sea on which it floats, so the main force to be overcome is resistance in moving through the water. That resistance is proportional to the area of the hull, a figure that can be expressed in, for example, square feet, while the value available inside the hull would be in cubic feet. Grossly simplifying the problem, if you took a starting unit of 2 units, then the area is 4 units, the volume 8 units: double that to 4 units and the area is now 16, but the volume is 64: the ratio has changed from twice as much to four times as much. The answer was to build big, and that was what Brunel recognised and what he proposed to do.

The ship, *Great Western*, was built by William Patterson of Bristol. They were not the only ones preparing for an Atlantic steamer crossing.

The British and American Steam Navigation Co. had placed an order for a paddle steamer, but there were problems, and in a bid to be first they purchased an existing vessel, the *Sirius*. The difference between the two contenders was considerable: *Sirius* was rated at 700 tons and carried forty passengers; *Great Western* was 1,340 tons and had first-class accommodation for 128 passengers with extra space for servants. The Brunel ship carried sail and had to make a voyage down to London to be fitted with her engine. While Brunel was on board for its sea trials, he was surprised to find *Sirius* steaming off for Cork to start her voyage. *Great Western* set off shortly afterwards for Bristol, but had scarcely got under way before a fire broke out in the forward boiler room. Caxton and others started dowsing the blaze and Brunel decided to see for himself. He was just starting down the ladder when a rung gave way and he fell 18ft. He would almost certainly have been killed, but he landed on top of Claxton. The ship was laid up to allow the injured engineer to be taken ashore before leaving for Bristol. On 8 April 1837, she finally set off for New York, but *Sirius* was already on her way, having left from Cork for the shorter crossing on the 4th. It was too big a gap to make up, but *Great Western* gained on her smaller rival all the way. *Sirius* arrived at New York on 23 April and *Great Western* a few hours later, but significantly the smaller vessel had used up all her coal – there were even rumours that cabin doors had been removed and burned – while the Brunel ship still had 200 tons in her bunkers. Big really was best.

Brunel appreciated that the future of ocean travel lay with big steamers, but he had approached the practical limits for wooden-hulled vessels. The announcement that a new vessel, *City of New York*, was to be built was made in September 1838. Initially it was intended to be wood as well, but with an overall length of over 250ft and breadth of 40ft it was pushing the limits. Brunel took a momentous decision: the new ship would have an iron hull. It was not the first iron ship, but nothing on this scale had ever been built. Work began at the Patterson yard in Bristol in July 1839. Like its predecessor, it was to be a paddle steamer and the job of designing the engine went to a young engineer, Francis Humphreys. He had earlier built an engine at the Hall works in Dartford, the company at which the pioneer steam engineer Richard Trevithick had spent his last days. There was a technical problem, however: no one was sure how to forge the immense crank shaft that was required. The problem was put to James Nasmyth,

9.3 The screw steamer *Great Britain* in heavy seas. (The SS Great Britain Project)

who promptly sat down and worked out a design for a steam hammer that could do the job. Things seemed all set when another inventor's work came to Brunel's attention.

Francis Pettit Smith had been working on a new type of propulsion. As the name suggests, the Archimedean screw had been around for over 2,000 years. It consists of a helical screw in a tube. If one end of the tube is submerged in water and the screw turned, then water will be lifted up the tube. Smith recognised that the screw could also be used to push water away as well as pull it up, and if it was then attached to a boat, as the water was thrust back, the boat would move forward. Smith tried various types of screw on model boats, before eventually discovering than a single turn was all that was necessary. It was to be known as a propellor. He made a full-scale version that was tried out on a launch, then went to an even bigger propeller on the schooner *Archimedes*. Guppy went on the first voyage of the ship and was so impressed that he persuaded his company to hire the vessel for a six-month trial. At the end, Brunel was convinced: the iron paddle steamer idea was scrapped. The new vessel would be a screw steamer. It was excellent news for the shipbuilders, but devastating for young Humphreys, who died shortly afterwards.

The ship, renamed the *Great Britain*, left Bristol in December 1845 for her sea trials, which proved a great success. Everything seemed to be going well and she began her scheduled voyages to New York. In September 1846 she set off again from Liverpool for New York. Her course should have taken her south of the Isle of Man, where she would have turned north to round the Irish coast to head out into the Atlantic. But due to a major navigational error, the captain failed to alter course, ploughed straight on past the Calf of Man and ran aground in Dundrum Bay in County Down. Brunel went across to Ireland and was horrified to find the great ship still stranded and unprotected from the elements. He took charge of plans for protecting her from the winter weather and eventually in August of the following year was able to get her refloated and towed back to Liverpool for repairs. It had been an expensive operation and the ship had been insured for only a fraction of its true value. As a result, the vessel that had cost almost £120,000 had to be sold off for a niggardly £18,000. The story does, of course, have a happy ending. The vessel was eventually brought back to Bristol and the dock where she was built, and has now been beautifully restored.

During this time, when Brunel was busy with ships and shipbuilding, he had many other projects in hand. In 1840, he was commissioned to design a bridge across the Thames at Hungerford. This at least gave him time in London with his family, which now included a young Isambard born in 1837. Two other children followed: Henry Marc in 1842 and Florence, whose birth was unrecorded, but was probably in 1847. The bridge itself was a suspension type that was destined to have a short existence. It was dismantled in 1860 to make way for the railway and pedestrian bridge that now occupies the site. The suspension chains, however, were saved and sent down to Bristol, where they were used in the completion of the Clifton bridge.

At the same time, the GWR was extending its broad-gauge territory, with an ambition to provide tracks all the way to Birmingham. The company achieved the first stage by converting the old Bristol and Gloucestershire Railway and had every hope of buying up the line from Gloucester to Cheltenham. But it was outbid by the Midland Railway, and, as a result, Brunel gauge and Stephenson gauge met head on at Gloucester. This caused a great deal of confusion in everything having to be changed – from passengers with their luggage to all kinds of freight. A Royal Commission was set up to decide what to do about it. After three months of hearing evidence, the GWR suggested a trial to see which was the better system. Stephenson would run one of his locomotives and a train on the level track between York and Darlington and Brunel would run his entry on an equivalent route from London to Didcot. This was agreed. Robert Stephenson chose *Ixion*, one of his new 4-2-0 long-boiler engines, and Brunel and Gooch put forward one of a new Firefly class. The Stephenson engine had the fastest time, achieving 60mph, while the Brunel only managed 53mph, but Brunel claimed that his engine was more economical and offered a more comfortable ride. In reality, the trial proved nothing, although one factor stood out: there was now a far greater length of the narrower gauge than there was of broad gauge. To convert the narrow gauge to broad gauge would be ruinously expensive, and every bridge, station and tunnel would have to be rebuilt to allow the extra width, while all that was needed to convert the broad gauge was to add an extra rail between those on the existing track. The decision was taken: no new broad-gauge lines would be authorised and, where necessary, the changes would be made to allow through routes where appropriate and necessary. The Stephenson gauge was now the standard gauge.

The GWR had not abandoned plans to reach Birmingham, this time through the Oxford, Worcester and Birmingham Railway, which appointed Brunel as its engineer. It should all have been straightforward, as there was only one major engineering feature, the mile-long tunnel at Mickleton. However, the line was to be responsible for one of the most controversial episodes of Brunel's career. Work on the tunnel was painfully slow: two sets of contractors started and gave up, and a third group led by Robert Marchant took over in December 1850. If Brunel hoped for better progress he was disappointed, and when his demand that more men be employed was ignored, the company decided to step in and take over the work itself. At this point, Marchant declared that he was owed £30,000 and barricaded the tunnel. He declared that he was prepared, with his navvies, to prevent anyone else working until he was paid. He quite cunningly went to the local magistrates to warn them that he might be attacked and got himself signed on as a special constable, although hardly an unbiased one. Brunel was incensed and decided to take his own considerably larger army of navvies to the site to force their way in. One of the magistrates, however, heard of the plan and was able to reach the site before any real trouble occurred. What became known as 'the battle of Mickleton' turned out to be a minor skirmish. Eventually the matter was settled in court. Marchant lost the case and was financially ruined, but Brunel was severely reprimanded for attempting to take the law into his own hands. Shortly afterwards the engineer heard that the company was no longer planning to make the route broad gauge after all, at which point he resigned.

In the middle of this hectic time in his life Brunel suffered an unusual personal problem. He loved entertaining children, and in April 1843 was doing a party trick with a gold sovereign that he accidentally swallowed. At first, he seemed untroubled, but after two weeks he noticed blood appearing and it seemed the coin was stuck in his bronchia. Brunel devised a scheme for dislodging it. He had himself strapped to a board that could be swivelled, until he was at 80 degrees to the ground with his head pointing down. He was then slapped on the back, but nothing appeared. His surgeon then decided he needed to open Brunel's chest and try to retrieve it but failed, so the two techniques were tried in combination. This time the coin popped up into his mouth and was spat out.

Brunel was always interested in new ideas, some of which proved entirely successful, others less so. One new idea was an electric telegraph system developed by William Coke and Charles Wheatstone, in which messages could be transmitted down a wire to a five-needle array that pointed to letters that could spell out words. They showed it to the London & Birmingham Railway, who showed no interest, and then took it to the GWR, where it was first put into use to connect signal boxes at West Drayton and Paddington in 1838. The world at large showed little interest until 1845, when there was a murder in Slough. The culprit was seen boarding a London train, a description was sent by telegraph and the police were waiting when he arrived. Ultimately the telegraph was to play a vital role in rail safety with the introduction of the block signalling system. When a train passed a signal box, the message was sent to the next box down the line, and no train was allowed onto that section until the message came back that it had cleared the second signal box.

The next novelty to attract Brunel's interest was the atmospheric railway. The basis is exactly the same as that of the Newcomen engine:

if you create a vacuum at one side of a piston, air pressure will move it. Here, however, the tube or cylinder is horizontal, with a slit on top covered by a leather flap weighted down with metal. A flange on the piston passes up through the slot and is attached to a vehicle running on tracks above. A steam engine is used to pump out air and, as the piston moves, it pushes the flap aside, which then falls back into place behind it. The first experimental section was demonstrated by Samuel Clegg on a 1½ mile length of track in south London. The idea was taken up by the engineers of the Dublin & Kingston Railway, where there was a very steep section of track. Their idea was to use the atmospheric railway to lift vehicles up the slope, which could then travel back down again by gravity. Brunel went to Ireland to see the system at work and was impressed. He was in the process of extending the broad gauge further west with the South Devon Railway, linking Exeter and Plymouth.

9.4 The atmospheric railway. (Nicholas Candy)

There were three steep sections, including the 3-mile long Dainton bank, which had a maximum gradient of 1 in 57. He decided to lay the whole line as an atmospheric system, with pumping engines at roughly 3-mile intervals. By 1846 the system was at work between Exeter and Teignmouth and at first all seemed to go well, until the winter brought cold spells that froze the leather valve. It had to be thawed out before any train could run. It then became clear that the leather did not survive well, and that it would need regular replacement. That would not only be expensive, but services would have to be halted while the work was done. There was no avoiding the brutal truth: it was a failure. It was dismantled and in June 1847 the service reopened with conventional locomotives. It was one of Brunel's biggest and most costly mistakes.

Another project attracted his interest: the plans for a grand exhibition hall in London, described in the previous chapter. Hundreds sent in their ideas, and the winner was Brunel. His building was a massive, forbidding brick structure, 2,000ft long and 500ft wide, topped by an iron dome. But as soon as his design was made public there was a storm of opposition. When the Crystal Palace was built to a rival design and the Great Exhibition opened, at least Brunel had the satisfaction of seeing the GWR locomotive *Lord of the Isles* given pride of place in the locomotive section.

Three years later, Britain went to war with Russia and troops were sent to the Crimea, only to be bogged down, literally and metaphorically, at Sevastopol. Brunel offered a design for a floating siege gun to the Admiralty. It could be towed into place by a special steamer, whose bows would open up to allow it to float out – a forerunner of the landing craft. The hull could be moved to bring the gun to bear using steam jets. Not only did the Admiralty fail to respond, but they actually lost the model he sent. Brunel was disgusted, but not surprised: 'They have an unlimited supply of some negative principle which seems to absorb and eliminate everything that approaches them.' Brunel was, however, to make a far more valuable contribution to the war. He was invited to design a prefabricated hospital that could be sent out to the Crimea and assembled on site. It was to have all the facilities needed, from well-ventilated wards to kitchen and laundry. The team went out and completed the entire project within seven weeks of landing. The old hospital at Scutari had been infamous for its wretched conditions, so bad that almost half of those admitted died not from the wounds, but from infections: of the 1,500 admitted to the Brunel hospital only forty died.

Brunel was still heavily involved in extending the GWR empire. Having failed to make the hoped-for progress into the English Midlands, the GWR began making plans to extend into South Wales. A new company was formed, the South Wales Railway, with Brunel as chief engineer. His greatest obstacle was the crossing of the Wye at Chepstow, where there was a high limestone cliff on the Welsh side and a requirement to keep the river open for boats. His solution was a hybrid. First, an embankment had to be built up on the English side, from which a conventional bridge on piers was built out to reach a quarter of the way across the river. A tower was built on top of the cliffs, and the final support in the river was extended upwards to reach the same height as the top of the tower. Hollow metal tubes were used to join the tops of the supports, from which metal ties were used to suspend the deck below. It was rather like a cross between a bow string and a suspension bridge. The bridge was opened in 1852 and trains could now run to Swansea.

The next line to be developed was the extension of the GWR line from Plymouth into Cornwall, which involved another complex river crossing, this time the Tamar. Brunel made use of the experience he had gained from the Chepstow bridge. This, however, was on a far greater scale, with an overall length of 2,187ft. There were to be seventeen conventional spans over the land, and two of 455ft each to cross the river. This time, instead of being horizontal, the iron tubes rose in two graceful arches. To construct the central pier, Brunel designed a caisson, 37ft in diameter and 80ft high, with the bottom shaped to fit the contours of the rock on the riverbed. The men worked inside the caisson and, once it had reached a height above high-water mark, the first of the great curved iron tubes was floated out and jacked into place. It was a complex process, over which Brunel presided like an orchestral conductor, but using flags instead of a baton. The bridge was completed in 1859 and Prince Albert arrived for the opening ceremony and named it the Royal Albert bridge. Brunel himself had been seriously ill, and crossed the bridge lying on a couch mounted on a flat truck.

9.5 The Tamar bridge under construction. (Walter Dendy)

With the line completed all the way from London to Penzance, it was time to replace the old Paddington station with something altogether grander. The design of the new Paddington was down to Brunel, who provided the overall pattern, while the architectural detail was the work of Matthew Digby Wyatt. Paddington remains one of Britain's great railway stations, and fittingly now has a statue of Brunel on the concourse, although the engineer might be surprised to find that he now shares the honour with a bear in a duffle coat.

The completion of the line to Penzance was to be Brunel's last great railway project, but he did have one other major scheme in mind. Having conquered the Atlantic, he began planning for a ship on a scale never built before that would be capable of a voyage from Britain to Australia. It was to be by far the largest ship ever built at that time and was to be driven by both paddle wheels and a propeller. In his initial notes, Brunel was considering a vessel 600ft long and 65ft wide. To put that in perspective, the equivalent figures for *Great Britain* were 322ft by 50ft. The vessel was ordered for the Eastern Steam Navigation Co. and, as it was far too large to be built in Bristol, the contract went to John Scott Russell, who had a yard on the Thames. Russell was initially enthusiastic, and so confident that when Brunel came up with an estimate for building the vessel and arrived at a figure of £275,000, with separate estimates for engines and boilers, Scott Russell offered to build it for £258,000. It was a decision he was to regret.

Brunel made it clear from the start that he was in charge of all design elements and Russell's contract included paying for the launch. There were soon disagreements between the two engineers. When Brunel made changes to the design, Russell began to ask for more money. He was then alarmed to hear that the launch, which he had agreed to pay for, would not be conventional, but would be one where the hull was to slide sideways down a ramp on specially built cradles, instead of bow first. Russell was bankrupt long before the launch day arrived. The day itself went badly when the great ship stuck on its launch rails, and there was an accident when one of the winches that controlled the movement was set free, throwing off one of the men, who died from his injuries. The vessel,

9.6 The second Paddington station, designed by Brunel and Digby Wyatt. (*Illustrated London News*)

9.7 The *Great Eastern* in New York harbour. (George Stacey)

now named *Great Eastern*, finally slid into the Thames on 31 January 1858. The stress of the launch and the arguments with Russell had their effect on Brunel's health. He had been diagnosed with a kidney disease, and he and his wife and son Henry set off across Europe and then on to Egypt to recuperate. They were joined at Christmas by Robert Stephenson. The following year there was the business of completing the *Great Eastern* to be overseen, and on 5 September 1859 he went on board to prepare for the sea trials due to start two days later. He had just arrived when he collapsed with a stroke that left him partly paralysed and he died on the 15th.

9.8 Clifton suspension bridge, completed after Brunel's death. (From *Lippincott's Magazine* 1878)

With his mercurial personality, his defiance of established views and willingness to try what others considered impossible, it is easy to see why Brunel is such a popular figure in British history. Yet it was his love of novelty and difference that led him to errors such as the disastrous atmospheric railway. To many he is best known for his work as a railway engineer, but the broad gauge was ultimately not the success he had hoped, and his ideas for locomotive design actually limited development in the first period of operation. The one area where he can be said to have succeeded not just in achieving the goals he had set for himself, but in transforming an industry was in shipbuilding: there his true greatness lies.

The ten engineers whose lives have been described here revolutionised transport and industry, not just in Britain, but throughout much of the world. Each had achievements of their own to celebrate, but the achievements of the individuals were often based firmly on the work of their predecessors. There will be no league table of greatness here, but simply an acceptance that all contributed to one of the greatest revolutions in history: the Industrial Revolution that created a new world.

BIBLIOGRAPHY AND FURTHER READING

Bailey, Michael R., *Robert Stephenson*, 2003.

Boucher, George, *James Brindley, Engineer*, 1968.

Burton, Anthony, *George and Robert Stephenson*, 2020.

Burton, Anthony, *Marc and Isambard Brunel*, 2022.

Burton, Anthony, *Richard Trevithick*, 2000.

Burton, Anthony, *Thomas Telford*, 2015.

Hadfield, Charles, and Skempton, A.W., *William Jessop, Engineer,* 1979.

Hills, Richard L., *James Watt* (3 volumes), 2002.

Lead, Peter, *Agents of Revolution: John and Thomas Gilbert – Entrepreneurs*, 1990.

Metcalf, John, *The Life of John Metcalf*, 1795.

Rennie, Sir John, *Autobiography of Sir John Rennie, F.R.S., Past President of the Institution of Civil Engineers*, 1875.

Richardson, Christine, *James Brindley*, 2005.

Rolt, L.T.C., *George and Robert Stephenson*, 1984.

Rolt, L.T.C., *Isambard Kingdom Brunel*, 1957.

Rolt, L.T.C., *Thomas Telford*, 1958.

Skempton, A.W., *John Smeaton FRS*, 1991.

Smeaton, J., *Reports of the Late John Smeaton*, 1837 (republished 2011).

Smiles, Samuel, *Lives of the Engineers* (5 vols), 1862.

Telford, Thomas, *The Life of Thomas Telford, Civil Engineer*, 1838.

Trevithick, Francis, *The Life of Richard Trevithick*, 1872.

INDEX